Framing Floors, Walls & Ceilings

TAUNTON'S

FOR PROS BY PROS ®

BUILDER-TESTED | CODE APPROVED

Framing Floors, Walls & Ceilings

EDITORS OF

Fine Homebuilding

The Taunton Press

The Taunton Press, Inc., 63 South Main Street, PO Box 5506, Newtown, CT 06470-5506
e-mail: tp@taunton.com

Editor: Christina Glennon
Copy editor: Seth Reichgott
Indexer: Jay Kreider
Interior design: Carol Singer
Layout: Jennifer Willman
Cover photographers: (Front cover): Justin Fink, courtesy of *Fine Homebuilding,* © *The Taunton Press, Inc.;* (Back cover): Patrick McCombe, courtesy of *Fine Homebuilding,* © The Taunton Press, Inc.

The following names/manufacturers appearing in *Framing Floors, Walls & Ceilings* are trademarks: Big Red®, Bosch®, Bostitch®, Dalluge®, Dandy Bag®, DeWalt®, Dumpster®, Duo-Fast®, Earth Saver®, Estwing®, Goo Gone®, Goof Off®, Grip-Rite®, Hitachi®, Honda®, iLevel®, Makita®, Max®, MemBrain™, Mr. Potato Head®, Occidental Leather®, Parallam®, Paslode®, Pneu Tools®, Proctor®, Senco®, Simpson Strong Tie®, Soil-Tek®, Speed Square®, Stanley®, Stiletto®, Terrafix®, Tremco®, TrusJoist®, Vaughan®, ZIP System®

Library of Congress Cataloging-in-Publication Data
Framing floors, walls & ceilings / author, editors of Fine Homebuilding.
    pages cm
 Includes index.
 ISBN 978-1-63186-005-8
1. House framing. I. Fine homebuilding.
 TH2301.F744 2015
 694'.2--dc23
                          2014047262

Printed in the United States of America

10 9 8 7 6 5 4 3 2 1

About Your Safety: Construction is inherently dangerous. Using hand or power tools improperly or ignoring safety practices can lead to permanent injury or even death. Don't try to perform operations you learn about here (or elsewhere) unless you're certain they are safe for you. If something about an operation doesn't feel right, don't do it. Look for another way. We want you to enjoy working on your home, so please keep safety foremost in your mind.

ACKNOWLEDGMENTS

Special thanks to the authors, editors, art directors, copy editors, and other staff members of *Fine Homebuilding* who contributed to the development of the articles in this book.

Contents

As soon as I finish this introduction, I'm going to walk into my backyard to frame the floor of a small shed that I'm building with 2×8 joists salvaged from a deck I recently took off my house. I haven't done any planning because the shed is going to be a simple 8-ft. square with a barn-style door (which I'm also building) and one big salvaged window. I'll frame the floor and walls on 16-in. centers. The walls will be overbuilt with 2×8 studs; I saved a lot of lumber from that deck, and I'm not interested in spending the time to rip the 2×8s down or engineering something that uses less material. If I were building or remodeling a house and paying for the materials, I wouldn't have this luxury. The framing would be most important. Not only would it be important for structural safety and for keeping costs in check, but the framing would also affect the finish and the work of all the trades that come after the framers. Framing is not as simple as it once was.

I'm not sure which came first: more complicated home designs or the tools and materials that we use to build them. They likely developed together, but gone are the days of framing a rectangular, gabled house with a tape, bubble level, hammer, and a stack of solid lumber. These days we use lasers, pneumatic tools, and a variety of man-made materials to frame great rooms, bump-outs, cantilevers, and lots of other awesome but complicated structural designs. Today more than ever, it's important to have the right tools, choose the best materials, and work as efficiently as possible.

That's where this book can help. Whether you are a seasoned builder or a novice carpenter, this collection of framing articles from the editors of *Fine Homebuilding* will deepen your understanding of framing as engineering, expand your horizon of tools and materials, and show you how to do the work fast and right. And once you've read through the basics, we'll show you how your framing can improve the energy efficiency of the homes you build. The framing is the bones of a house, and the bones are just as important to a house as they are to you and me.

Build well,

Brian Pontolilo, Editorial Director, *Fine Homebuilding*

On the Job Site

Choosing the Right Framing Nailer

BY MICHAEL SPRINGER

B uilders often ask for tool tests of framing nailers, but all the variations on the market make
that a tall order to fill. There are stick nailers
with 20-, 28-, and 30-degree magazine angles, not to
mention coil nailers. Some tools max out at 3¼-in.
nails, some at 3½-in. nails, and some at 4-in. nails
or longer. Some tools shoot full round-head nails,
clipped-head nails, or both. With the variety of
models available from the major pneumatic brands,
power-tool companies, and lower-cost clone and
private-label manufacturers, the framing-nailer
category must represent 100 or more tools.

Here, my goal is to condense all the relevant
information about these nailers into a brief guide,
highlighting the latest technologies and features
these tools have to offer.

It All Starts with the Nails

Picking a framing nailer starts with knowing the
nails you'll be shooting. You want a tool that you
can keep supplied with nails easily and affordably.
Regional preferences and sometimes even building
codes dictate which fasteners—and therefore which
tools—are common in your area.

California and other Western states have adopted
full round-head nailers, whereas most of the rest of
the country relies on clipped-head models. Specific

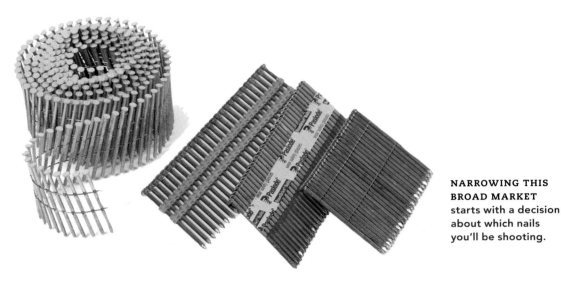

code requirements have driven some of the divide, but these geographic tool preferences can be traced back to where the big nailer companies started, or at least to the regional markets where their distribution was originally focused. Think Bostitch® in New England, Hitachi® in the West, and Paslode® and Senco® in between. As the major players staked their claims, whatever type of nail their early tools required became the default favorite in the territory.

Regardless of nail type, follow the nailing schedule for each material, component, and assembly you construct as specified by the building code covering your area. Model building codes were written for hand-driven nails, so they specify only the size, spacing, and number of nails used for specific connections and applications, not the type of head. The International Code Council's ESR-1539 report—which is free and widely available online—is written with an awareness of pneumatic nailers and is a good place to find the details of nailed connections (and equivalent connections) required to meet all the model building codes.

Nail Collation

Framing nailers come in two styles: coil or stick. Coil nailers have an adjustable canister that accepts a coil of nails strung together by two rows of thin wire welded to the shanks of the nails. These nails have a full round head. Stick nailers fit two angled sticks of 25 to 40 nails collated with wire, paper, or plastic, with the head of each nail nested just above the head of the nail in front of it. The style of nail head is usually based on the collation angle.

Nail Heads

Full round-head nails are acceptable everywhere in the United States and for every type of framing connection. They are also typically available in thicker shank diameters. The downside is that nail heads take up space in a magazine, so you get fewer nails per stick.

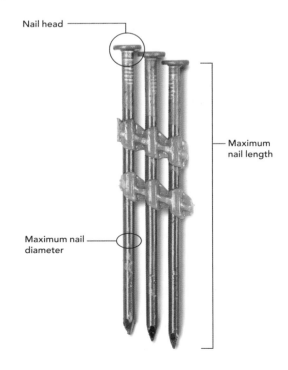

Nail head

Maximum nail length

Maximum nail diameter

Maximum Nail Diameter

Nailers also have limitations to the maximum diameter of compatible fasteners they can accept. The minimum nail thickness for wall sheathing isn't typically the same as the minimum nail thickness for rafters. This varies by region, though, and also can change based on what the architect or engineer has specified in the building plans. I know a framer in the Southwest who is allowed to use 3-in. by 0.131-in. (10d) nails for everything—an easy task for any framing nailer.

Maximum Nail Length

Some brands have created a new compact-framer category, designed to be lighter and to fit more easily between 16-in.-on-center framing layouts. The dividing line for this category is typically maximum nail length—3¼ in. for compact models, 3½ in. for full size—but the maximum shank thickness also may differ by collation angle and brand.

The maximum size is often referenced in a nailer's model number. For instance, a domestic model number may express maximum nail lengths of 3¼ in. and 3½ in. as 325 and 350. Foreign models may use 83 and 90, which are the lengths in millimeters.

It's worth noting that some compact nailers don't have the guts to shoot into dense engineered lumber well. Even if the longest nails you shoot are 3¼ in., you may be better off with a full-size nailer because of its superior power.

Must-Have Features

Some other important features to consider when choosing a nailer are:

Balance and feel are important to your overall comfort and the control of the tool. Before you plunk down cash, be sure to fill the tool with nails and to hang a hose off the back to evaluate how it really feels. Otherwise, you're just kicking the tires.

The body of the nailer will be either aluminum or magnesium, and the choice is a bit of a toss-up. Magnesium is lighter but more brittle, and it costs more than standard aluminum, which is heavier and

LOADING OPTIONS. You'll need to decide if you want a rear-loading magazine, like the one shown, or a top-loading magazine.

more durable. It's best just to go with how the tool feels overall, though. I don't know that anyone buys a nailer based on the material it's cast from.

A selective-fire setting lets you switch the tool from sequential-fire (single-shot) mode to bump-fire mode. The best designs are tool free, but because most users never switch back to sequential fire, replacing or adjusting the trigger assembly once is not a big deal. If you plan to switch back and forth, opt for a nailer that has a toggle switch.

Top-load versus rear-load magazines is a decision you will have to make. For myself and the guys I know, the answer is unanimously in favor of rear load. Hanging the tool down with one hand lets you load in a more comfortable position; the spring-loaded follower can't accidentally slam into the nails and damage the collation strip; the remaining nails can't fall out as soon as you release the follower; and having the follower engaged when loading new nails keeps the last few remaining nails tightly in place so that they won't cause a jam.

Depth-of-drive adjustment is important for meeting building codes, and the best setups work without the need for tools. Without this feature, you have to adjust the regulator on the air compressor when you switch from LVL headers to nailing off sheathing. It's not worth compromising on this feature.

FRAMING NAILERS: COIL OR STICK?

COIL NAILERS

COIL NAILERS HAVE AN ADJUSTABLE CANISTER THAT accepts a coil of nails—up to 200 framing nails or 300 sheathing nails at a time—angled at 15 degrees and strung together by two rows of thin wire welded to the shanks of the nails. In most areas of the United States these nailers are far less popular than stick nailers, but they are common in areas of the Northeast and in a few pockets of Louisiana, Missouri, and Texas. Interestingly, this is what the rest of the world considers a framing nailer.

PROS

- These tools shoot a lot of nails between reloading, potentially saving time.
- The tools' compact size provides some accessibility advantages.
- If a model fits shorter nails and has a protective nosepiece, it can double as a high-volume siding or trim nailer.

CONS

- When fully loaded with hundreds of nails, these tools can be heavy and unwieldy.
- Dropping or bending a coil of nails often renders it unusable and creates expensive waste.

NOTABLE BRANDS
Bostitch, DeWalt®, Grip-Rite®, Hitachi, Makita®, Max®, Pneu Tools®, Senco

STICK NAILERS

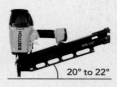

20° to 22°

PLASTIC-COLLATED NAILERS fit round-head nails collated between 20 degrees and 22 degrees. A stiff collating strip—typically plastic but also available in rigid paper—allows enough space for full-size heads with the nails situated side-by-side. Two sticks of nails fit in the magazine for a load of about 60 nails. Full round-head nails have been a necessity in some parts of California for a while, so these tools are particularly big on the West Coast and in much of the West in general.

PROS

- Round-head nails are allowed for every connection type, so these tools can be used anywhere in the United States with their standard fasteners. (Some codes require the use of round-head nails only.)
- Round-head nails are typically available in larger shank sizes than other types.
- The easier manufacturing of plastic-collated nail sticks makes them significantly less expensive than the paper- or wire-collated nails used in other nailers.

CONS

- Nails of the standard plastic-collated variety spew out bits of plastic shrapnel, which is a nuisance when they ricochet off the wall into your face or leave the floor dotted with scattered shards.
- The long, low magazine keeps the nose of the tool from fitting into tight spots as easily as higher-angle stick nailers.

NOTABLE BRANDS
Plastic-collated:
Bosch®, Bostitch, DeWalt, Duo-Fast®, Grip-Rite, Hitachi, Makita, Max, Pneu Tools, Senco.
Wire-weld-collated:
Bostitch, Grip-Rite, Hitachi, Max. Paper-collated: Bosch, Bostitch, DeWalt, Grip-Rite, Hitachi, Max, Paslode, Pneu Tools, Senco.

Two Head-Style Options for Paper-Collated Stick Nails

Depending on your region and applicable codes, the type of head on your nails is a big deal, and the head style is usually tied to the collation angle. There are a few variations.

OFFSET ROUND HEAD

Available in both 28-degree and 30-degree angles, these nails provide the code compliance of a full round-head nail with the tight spacing common to a stick of clipped-head nails.

Offset round head

CLIPPED HEAD

At steeper collation angles (28 degrees and 30 degrees), manufacturers can pack nails closer together by clipping off one side of the head. The resulting D-shape has less surface area than round heads of the same diameter and causes these nails to be disallowed for some applications. Some of the nails are called notched instead of clipped because the chunk removed from their heads is rounded instead of straight.

Clipped head

WIRE-WELD-COLLATED NAILERS have their own specific collation angle and their own specific homegrown market. These tools started strong in the Northeast and have stayed strong, with 80 percent of their sales in New England. Overall, the 28-degree tools are similar to the 30-degree type (they share the same pros and cons), but 28-degree clipped-head nails are typically collated with thin wires tack-welded to the side of the nails. Plastic-collated and paper-tape versions of 28-degree nails can be found, even some with round heads, but before you bring one out to build shear walls in California make sure you can get fasteners for it. These tools are largely unknown in much of the country, and their diet of special fasteners may not be on the menu far from home.

PAPER-COLLATED NAILERS have magazine angles anywhere from 30 degrees to 35 degrees, but the fasteners they fit are usually referred to as 30-degree nails. These tools are known generically as clipped-head nailers or paper-tape nailers. Standard clipped-head nails for these tools are collated with paper tape glued along the sides of nails that are packed shank to shank. Magazines typically fit two sticks of these densely packed nails, providing about 80 nails.

PROS
- The steep magazine angle of the tool affords its nose the deepest reach into corners.
- Tighter packed sticks of nails hold significantly more fasteners per load than the sticks in full round-head nailers.

CONS
- Clipped-head nails are not approved for structural connections in some areas, so more specialized offset round-head nails may be needed.
- These tools require more expensive fasteners than full round-head nailers.

AVOID DAMAGE. A nonmarring nose cap keeps your wood safe.

HANDY UPGRADES

Whether you're framing on a regular basis or just looking for some extra perks, try these features.

A few manufacturers are using **nose magnets** to hold the last few nails in a stick firmly in place when reloading the magazine. This is a simple, useful addition to help make the reloading process goof proof.

Built-in air filters are a welcome addition to keep unwanted gunk out of a nailer's innards. Pads of filter media are useful enough, but the best filters are self-cleaning cartridges that cough out any trapped particles every time you unplug the air hose.

A **nonmarring nose cap** allows your framing nailer to become a siding, trim, or deck nailer. Without a cap, the teeth on the nose turn cedar or redwood into hamburger.

FEATURES FOR PRODUCTION FRAMERS

- **A RAFTER HOOK** for keeping your nailer close at hand is a must if you don't want your tool to slow you down. The lack of a hook is not a deal breaker, however. Aftermarket hooks that connect at the air fitting are available and may be preferable to a factory-installed hook that is clumsy or too small.
- **TOENAILING SPIKES** (see photo at left) that really grab are the key to fast, accurate toenailing. Look for a nose with especially sharp spikes protruding well out from the sides.
- **REAL-WORLD DURABILITY AND LONGEVITY** are hard to test. Many top brands have legacy tools that have been in production for decades, so talk to other guys in the field. Check local tool-repair shops, too. They might not be able to tell you which nailer will last, but they sure can tell you which ones break.

Hammers: Titanium vs. Steel

BY ROB YAGID

A few years ago, I worked on a framing crew run by a boss who took everything he did to the limit: building houses, racing dirt bikes, even buying tools. The day I walked onto the job, he handed me a leather Occidental Leather® tool bag that held a 16-oz. titanium Stiletto® hammer. I strapped on the belt and, to his surprise, replaced the hammer with a 22-oz. steel Estwing®. At the time, I didn't get the hype surrounding titanium. After much persuasion, though, I switched. I quickly fell in love with that Stiletto, but never understood why.

The hammer felt great, but how much could titanium really affect the tool? I wasn't sure if I loved that hammer for the way it performed or if I was just a sucker for savvy marketing. It turns out that the hype has value.

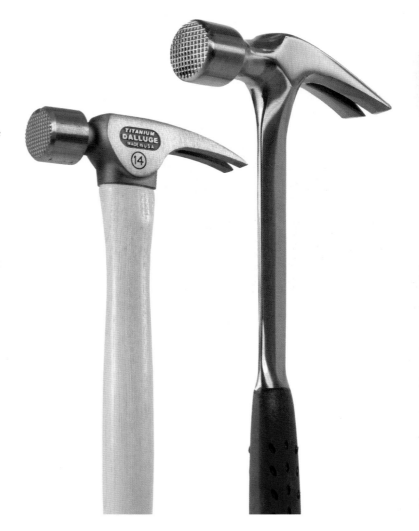

Titanium Transfers Energy Efficiently

Brandon Miller, the product marketing coordinator for Stiletto tools, sold me on titanium in two sentences. A titanium hammer transfers 97 percent of your energy from swinging the hammer to the nail, whereas a steel hammer transfers only 70 percent of your energy to the nail. That means titanium drives the nail more efficiently and there is less recoil energy to travel back into your arm (3 percent vs. roughly 30 percent with steel). This clarifies the claims that titanium is roughly 45 percent lighter than steel but hits just as hard, and that titanium can help to save your elbow from the shock of hammering.

The chief complaint with these hammers is their expense. In its raw form, titanium is roughly five times more expensive than steel. The higher cost to produce the tool contributes to a price tag of $80 to $250, depending on the hammer.

Steel Is a Deal and Still Packs a Punch

Although a titanium hammer might be more efficient at driving nails, some users prefer steel hammers for their knockdown power. Because steel is much heavier than titanium, it's more useful when it comes to pounding a beam into place, moving a wall into its proper position, or doing demolition work.

A steel hammer's most alluring quality, though, is that it's considerably less expensive than a titanium hammer. At around $20 to $30, it's the most economical choice for those who don't swing a hammer for a living and are not as susceptible to the physical damage that can occur from years of job-site abuse.

SOURCES

TITANIUM HAMMERS

DALLUGE® TOOLS
www.dallugetools.com

STILETTO TOOLS
www.stilettotools.com

VAUGHAN®
www.vaughanmfg.com

STEEL HAMMERS

DEAD ON TOOLS
www.deadontools.com

ESTWING
www.estwing.com

STANLEY® TOOLS
www.stanleytools.com

Framing Lumber: Moisture Content, Species, and Grade

BY DON BURGARD

ngineers, architects, and builders need to know that the framing lumber they specify and install is strong enough to handle the loads placed on it. At the same time, they want to build cost-effectively, which involves knowing when a more expensive species or grade of lumber is un-necessary. For these reasons, each piece of framing lumber has a stamp that contains several pieces of information, three of which—moisture content, species, and grade—are essential to know before you start building.

Grade

Species

Moisture content

Moisture Content

Moisture content is identified by one of many abbreviations: AD (air-dried to a moisture level at or below 19 percent); S-DRY (surfaced dry; that is, the board when surfaced at the mill had a moisture level at or below 19 percent); S-GRN (surfaced green; that is, the moisture level when the board was surfaced was higher than 19 percent); KD (kiln-dried at or below 19 percent); and KD-HT (same as KD, although heat-treated as well to kill pests and fungi). KD-15 and MC-15 lumber have moisture levels of 15 percent or less.

Some builders choose lumber with a higher moisture level for new construction. It's cheaper, it's less prone to splitting when nailed, and all the wood will shrink at a similar pace as it dries. Others prefer drier lumber to avoid problems related to shrinkage, such as twisting and nail pops. For remodels, however, go with the drier stuff because it will integrate better with the already dried and shrunken existing lumber. Be aware, however, that moisture content is measured at the mill, not at the lumberyard, so time sitting in rainy weather or hot sun isn't considered.

Species

Common wood-species stamps include D Fir (Douglas fir), Hem (hemlock), and PP (Ponderosa pine). Abbreviations are sometimes grouped for species with similar characteristics, such as SPF (spruce, pine, and fir) and Hem-Fir (hemlock and fir).

Species can depend on region, but high-strength lumber often is more expensive. That said, it isn't always necessary to use the strongest lumber available. Building codes specify maximum allowable spans for each species, so before buying expensive lumber, check local codes to see if a cheaper species could work.

Grade

There are four categories of framing lumber, most of which have a hierarchy of grades corresponding to their level of weakening characteristics, such as knots, splits, or wane.

Structural light framing: These pieces have the highest strength values and are suitable for use in engineered applications such as trusses, rafters, and joists. They are broken down into four subcategories: select structural, No. 1, No. 2, and No. 3.

Light framing: These pieces can be used as plates, cripples, blocking, and in other areas where high strength isn't crucial. They break down as construction, standard, and utility.

Stud: These pieces are strong enough to handle vertical loads, but they aren't approved for other uses. No subcategories here.

Structural joists and planks: These larger boards have the same grades as structural light framing.

The farther down the hierarchy you go for each category of lumber, the lower the quality and performance of the wood. Because different species have different strength values, grade must be considered hand-in-hand with species. For example, a 10-ft. Douglas-fir 2×4 of a certain grade will have a different strength value than a Ponderosa-pine 2×4 of the same size and grade. The most important thing to remember here is that within the same species you can always use a piece of lumber graded above what is required for a particular application, but not one that is graded below.

Choosing the Right Nail for the Connection

BY DEBORAH JUDGE SILBER

Considering how critical nails are to holding wood-frame houses together, it's surprising that we don't pay more attention to them. The fact is, nails and the connections they make are critical to managing building loads and ensuring a safe, durable structure. Although there are many types of nailed connections used in home building, the three basic wood-to-wood connections here illustrate how nails perform.

In service, nails must resist withdrawal forces and shear (lateral) forces; they must also be resistant to pull-through and to combined (off-axis) forces. How well they perform is dependent on the characteristics of the nail, the wood, and the angle at which the nail is driven. Altering these factors—such as using a ring-shank nail or driving the nail at an angle—has a much greater effect on withdrawal resistance than on shear resistance, which is more dependent on the bending strength of the nail and the bearing capacity of the wood surrounding it. Penetration is also important. The rule of thumb is that at least two-thirds of the nail should extend into the base material. So a 1×3 should be fastened to a 4×4 with a 2½-in. (8d) nail, with ¾ in. of the nail going through the 1×3 and 1¾ in. going into the 4×4.

So the next time you swing a hammer, consider this: There's a lot riding on that nail. Here's how it works.

THREE BASIC CONNECTIONS

END-NAILED CONNECTIONS

This connection joins two wood members whose grain direction is perpendicular. These connections are easy to make, but they offer little withdrawal resistance (up to 75% less than a nail driven perpendicular to the grain) and effectively resist only shear (lateral) forces.

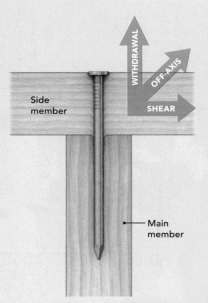

Side member

Main member

FACE-NAILED CONNECTIONS

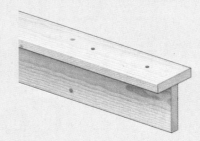

This connection joins wood members with the grains parallel. In this application, the nails resist withdrawal, shear, and sometimes off-axis forces.

The bottom of the head should press on the side member but not be driven deeper than the head thickness.

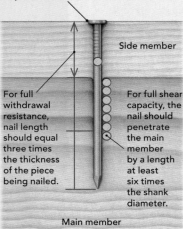

Side member

For full withdrawal resistance, nail length should equal three times the thickness of the piece being nailed.

For full shear capacity, the nail should penetrate the main member by a length at least six times the shank diameter.

Main member

TOENAILED CONNECTIONS

This connection offers both withdrawal and shear resistance regardless of the grain direction of the members being nailed. Tests show that these connections are made strongest by using the largest nail that will not cause splitting, by inserting the nail one-third of its length from the joint, by driving the nail at a 30-degree angle, and by burying the full shank of the nail without causing excessive damage to the wood. When driving several nails, cross-slant driving is somewhat more effective than driving the nails parallel.

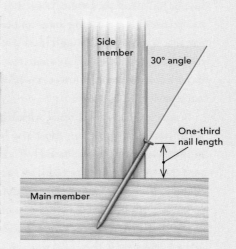

Side member

30° angle

One-third nail length

Main member

OTHER COMMON CONNECTIONS

DOVETAIL NAILING
Nails are driven at an angle through the face of a board to clamp the boards together and to provide better withdrawal resistance than perpendicular face-nailing.

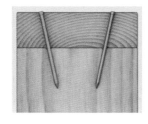

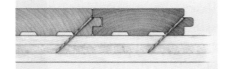

BLIND NAILING
This connection is used with tongue-and-groove boards. Nails are driven at a 45-degree angle, enabling the groove of the adjacent board to fit over the nail.

CLINCH NAILING
An extralong nail is driven through the wood members being joined, and the tip is then bent and nailed flush for extra withdrawal resistance.

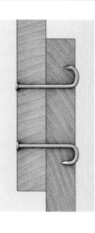

ANATOMY OF A NAIL

HEAD
Head size and structure vary with a nail's type and purpose: to avoid overpenetration of the nailed material (such as the broad thin head of a roofing nail) or to embed the head in it (such as the barrel-shaped head of a finishing nail). Embossed nail heads enhance paint adhesion. Head shape has little bearing on withdrawal, but a small head can result in pull-through under force.

Finishing   Roofing

GAUGE/DIAMETER
Gauge is a measure of nailshank diameter most commonly associated with collated nails. The larger the diameter, the lower the gauge. In general, nails with a large diameter have greater resistance to withdrawal and lateral loads.

SHANK
Smooth-shank nails drive into and pull out of most woods more easily than deformed-shank nails, but those deformations—for example, annular rings or helical threads—can improve holding in certain materials such as hardwoods or plywoods. Ring-shank nails can have up to twice the withdrawal capacity of smooth-shank nails.

Annular   Helical

POINT
Most nails have a four-sided diamond point to make driving easier. Sharp points enhance withdrawal resistance, but they can cause wood to split. Blunt points prevent splitting but lessen withdrawal resistance.

SIZE
The size of nails for wood-to-wood applications is commonly referred to by pennyweight. The term is attributed to the original price per hundred nails and is designated with a "d" (for the Roman coin *denarius*). Pennyweight identifies nails by size on an established but somewhat arbitrary scale (below) and is not considered the best method of specification. Both shank length and diameter can vary slightly among different nails of the same pennyweight: For example, an 8d common nail measures 2½ in. with a 0.131-in. dia., an 8d box nail measures 2½ in. with a 0.113-in. dia., and an 8d sinker nail measures 2⅜ in. with an 0.113-in. dia.

Penny–inch nail equivalents (common)

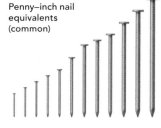

2d = 1 in.	10d = 3 in.
3d = 1¼ in.	12d = 3¼ in.
4d = 1½ in.	16d = 3½ in.
5d = 1¾ in.	20d = 4 in.
6d = 2 in.	30d = 4½ in.
7d = 2¼ in.	40d = 5 in.
8d = 2½ in.	
9d = 2¾ in.	

MATERIAL
Typically made from carbon-steel wire, nails also can be made from aluminum, brass, nickel, bronze, copper, and stainless steel. These materials have different friction values and bending strengths, influencing withdrawal and shear capacity.

COATING
Sacrificial galvanized (zinc) coatings delay corrosion of steel nails. Hot-dipped galvanized nails are immersed in molten zinc to produce a durable coating; other processes include mechanical galvanization and electrogalvanization. Polymer coatings increase initial withdrawal resistance by increasing friction between the nail and the wood. Driving into hardwoods can remove this coating, however.

Building Sturdy Sawhorses

BY PATRICK McCOMBE

I once had a coworker who described his favorite sawhorses as Clydesdales. It was an appropriate name, because his bulletproof horses seemed as heavy as their 1,000-lb. namesake.

Another coworker had the opposite approach. He favored lightweight sawhorses made from six equal-length pieces of 2×4. He could bang a pair together in the time it takes to roll out an extension cord.

I think the best sawhorses are a combination of these two approaches. The best design I've found is the easy-to-build "Stackable site-built sawhorses" from Ty Simmons of Fort Laramie, Wyoming.

I like Ty's horses because they're lightweight and strong, and use common materials.

You could build them with only a circular saw, but using a tablesaw for beveling the top and a sliding miter saw for cutting the legs makes the process much easier.

Including setup, it took me about 90 minutes and cost me $50 in materials to make the pair of horses featured here.

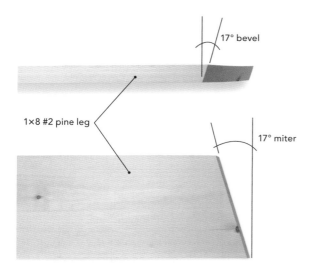

17° bevel

1×8 #2 pine leg

17° miter

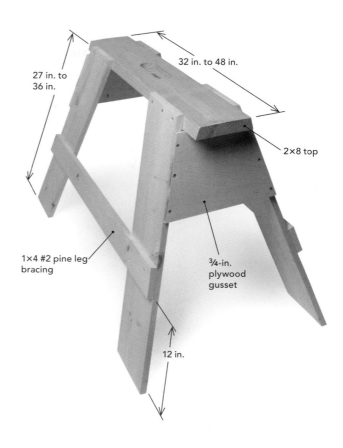

27 in. to 36 in.

32 in. to 48 in.

2×8 top

1×4 #2 pine leg bracing

¾-in. plywood gusset

12 in.

BEVEL THE TOP. Rip a 17-degree bevel on both sides of the top. You can do this with a circular saw, but a tablesaw with a rip fence is faster and easier.

CUT THE LEGS. All legs have a compound angle on both ends. With the miter and bevel both set at 17-degrees, cut four legs (for a pair of horses) so that the finished leg has the same cut on both ends. Leaving the bevel setting alone, rotate the miter table to the other 17-degree mark on the other side of 90 degrees, and cut the other four legs.

FASTEN THE LEGS. Mark the top 3 in. from the end at all four corners. Hold the leg flush to the top, and drill the pilot holes with a twist or pilot/countersink bit. Then drive the screws. Use 1½-in. #10 wood screws because they have greater shear strength than deck or drywall screws. Trim one of the tops later if you want the horses to stack.

INSTALL THE BRACING. Mark the legs 12 in. up to locate the 1×4 leg bracing. Hold or clamp the board in place, and scribe both ends to match the leg angle. Drill the pilot holes, and drive the screws.

SCRIBE AND CUT THE GUSSETS. Holding a scrap of plywood against the legs, scribe the sides of the gusset, and cut it with a circular saw. Clamping or screwing the plywood to the nearly completed sawhorse makes cutting safer and easier.

ALMOST DONE. Clamp or hold the gusset in place while you drill pilot holes. Be sure to align the drill bit carefully so that the hole stays in the center of the leg. After you run in the screws, your team of horses is ready for work.

FOUR MORE OPTIONS

CONSIDER THESE ALTERNATIVES WHEN space is tight or time is short.

SITE-BUILT

SPACE SAVER. Built from 20-in. by 32-in. scraps of ¾-in. plywood, these clever knockdown sawhorses are surprisingly sturdy. The ⅜-in.-deep mortises in the 2×6 top are made with a circular saw and a sharp chisel.

GET 'ER DONE. Made from six 32-in. pieces of 2×4, these are the horses to build when you're in a hurry. You can use gun nails if time is extra tight, but 2½-in. screws hold better and come apart easier when it's time for your steeds to become blocking.

STORE-BOUGHT

FAST, ROCK SOLID, BUT EXPENSIVE. Trojan's nearly indestructible TS series is the best commercially produced sawhorse, but you'll need to shell out $90 per pair for the 27-in. version. The 35-in. model costs another $12 per pair.

A POPULAR OPTION. Available at home centers everywhere, these sawhorses ($60 per pair) have self-leveling, adjustable legs that keep the horse steady on uneven terrain, but it's easy to lose the leg extensions and the bolts that secure them.

Exploring the Benefits of Engineered Floor Joists

BY CHRIS ERMIDES

I t used to be that architects and designers were limited by floor-framing choices; dimensional lumber was the only thing available. Depending on species and age, a 2×8, 2×10, or 2×12 might require supporting beams or bearing walls, both of which could limit design possibilities. Large, uninterrupted spaces were complicated to design and build.

Engineered floor-framing options such as wood I-joists and floor trusses, however, can span greater lengths with fewer caveats, yielding open spaces easily accessible to designers and framers. Both I-joists and floor trusses minimize or eliminate potential engineering problems and in the end—despite the added upfront cost—might even save money. Engineered wood joists and trusses are easier to specify, are more stable than dimensional lumber, and are easier to install.

Load Affects the Top and Bottom of Floor Joists

I-joists were developed to meet the demands of open floor plans. Architects needed a material that could clear-span areas larger than dimensional lumber without complicated engineering. When they arrived on the market in 1969, I-joists were made of a plywood web capped with top and bottom flanges. Those flanges were either laminated-veneer lumber (LVL), solid lumber, or strand lumber. Today those webs are made of oriented strand board (OSB), whereas the flange options are limited to LVLs and sawn lumber.

One advantage I-joists have over dimensional lumber is consistency and stability. Dimensional lumber can vary in width from board to board, and even end to end. It shrinks over time, and framers have to be mindful of splits, checks, twists, and crowns in every board. Setting a single joist becomes a multistep process. Also, the possibility of shrinkage means potential drywall cracks and uneven, wavy ceilings.

The most exploitable advantage I-joists have over dimensional lumber, though, is strength. At first glance, it's tough to imagine that a ⅜-in.-thick web of OSB capped with 2×2 blocks of wood could be as strong as a solid piece of lumber. To accept that fact, you need to understand how load stresses are distributed within an I-joist.

As a load bears down on an I-joist, the load stresses mainly the top and the bottom flanges. The wood fibers on the top flange are compressed, whereas the fibers on the bottom flange are stretched. The fibers near mid-depth, however, are virtually unaffected.

I-JOISTS

LVL flanged I-joist

Sawn-lumber flanged I-joist

I-JOISTS NOW MAKE UP MORE THAN 50 PERCENT OF THE floor-joist market. They're strong, lightweight, stable, and more versatile than dimensional lumber. Longer spans, varying depths, and a range of flange widths provide builders with options for dialing in price and performance. They're available from national and regional manufacturers through local lumberyards. Price varies by location.

OSB has replaced plywood, the original web material used by most manufacturers, each of which makes its own OSB webs to ensure the strength and quality of the joist. Proprietary OSB webbing is often designed to resist moisture, which can pose a serious threat to the integrity of the flange-to-web bond.

The flanges are made of either solid dimensional lumber or LVL material. LVL flanges are made up of thin wood veneers, which manufacturers claim means that they're more stable. Wooden flanges are as strong as most LVL flanges, but they are more susceptible to movement from expansion and contraction. It's important to note, however, that expansion and contraction are minimal because these flanges are attached to stable OSB webbing. Manufacturers of solid-wood flange types maintain that those flanges are as stable and resource-efficient as LVL flanges, and more economical as well.

PROS
- Available in many sizes and configurations so that performance can be somewhat customized.
- Lightweight so that they're easy to maneuver around the job site.
- Install like dimensional lumber, which most framers are familiar with.
- Span and load ratings are predetermined, so layout can be reconfigured on site as needed.
- Dimensionally consistent and stable.

CONS
- OSB-to-flange connection is susceptible to damage if exposed to excessive moisture.
- Not structurally stable until bracing and/or floor sheathing is installed.
- Web stiffeners are required at point loads and when hangers are used, which means more time to install.

FLOOR TRUSSES

Open-web floor trusses are gaining popularity, especially in custom homes. They're lightweight, strong, and capable of making long spans. Builders like them because they can be fully customized. For example, multiple live and dead loads can be designed along a set of trusses, eliminating the need for bearing walls or beams. It's also possible to create utility chases up to 24 in. wide at any desired location. Open webs make running wire and pipe easier, and having a dedicated chase makes work easier for subs. A truss designer can spec a set of trusses to desired spans, deflections, point loads, and bearing capacity while taking into account the materials that will be added later. The bottom line: Performance is highly customizable.

Trusses are available mostly through regional manufacturers and local lumberyards. Prices depend on design and location.

Stress-rated 2× pine is used for both top and bottom chords, as well as for webbing. The chords tend to run a higher grade, usually No. 2 or better. Web lumber is stress-rated as well, although it isn't usually as high grade. This is where manufacturers keep costs under control; higher-rated wood is used where it's needed according to the load design for a particular truss.

Web members join to top and bottom chords by heavy-gauge steel nailing plates. These plates come in various sizes and gauges that are pressed into the wood at each joint by a large roller or hydraulic press.

Top chord-bearing truss

Bottom chord-bearing truss

PROS

- Customized layout and truss design with an engineer's stamp means less time and worry for the builder.
- Customized performance means guaranteed results.
- Utilities are easy to run and are accessible later, saving time and money.
- Easy and fast to install.
- No additional bearing blocking is needed.
- Wide bearing surface makes for faster, more accurate subfloor installation.

CONS

- Customization means that layout changes can't be made in the field.
- All layout/structural changes have to be approved by an engineer and might require new trusses.
- Framers unfamiliar with them can install them improperly: upside down, front to back, or in the wrong order.

This is why it's OK to drill holes along the center—and also why you can't notch the top and bottom of any joists. Notches remove wood fiber from the joist areas that need it most.

Builders such as Michael Chandler in Raleigh, North Carolina, like I-joists for their strength and value, but also for their sustainable qualities. "My goal is to build houses that will last more than 100 years," Chandler says, "and I want to do so with minimal impact on the environment." The fact that I-joists use far less wood than dimensional lumber is a big benefit for many builders.

I-Joists Offer Builders Flexibility

I-joist manufacturers have figured out a way to optimize the physics involved in the way bending stresses are distributed within a joist, and they have designed a product that puts wood where it's needed most and removes it where it's needed least. The result is a stronger, straighter, and lighter joist than dimensional lumber.

Save for a few necessary stiffeners, I-joists install similarly to dimensional lumber, and they offer builders similar flexibility to dimensional joists. Because allowable spans are set by I-joist manufacturers, a framer can move a layout as necessary if changes need to be made after the joists are delivered. I-joists can be doubled for added strength or moved to make room for plumbing and other mechanicals. I-joists also offer more options for drilling holes.

Manufacturers such as iLevel® offer free software to help designers and builders customize performance. To help builders achieve desired floor performance, most manufacturers offer a selection of I-joists with variations in flange size, web thickness, and depth.

Floor Trusses Are Custom Joists

Given their strength, their efficient use of materials, and their open-web design, wood floor trusses are sustainably built and offer benefits to builders, subs, and even homeowners. Truss manufacturers can take a set of plans and customize the truss designs down to details of concentrated load placement and utility chases (see pp. 26–27).

Aside from their design characteristics, trusses are easy to install—as long as framers follow the truss manufacturer's placement diagrams. Lisa Biggin, construction manager for Habitat for Humanity in Newburgh, New York, likes trusses for several reasons. "I need to clear-span distances that would require structural supports for I-joists or dimensional lumber," she says. "The added cost for those supports and columns would make the floor system more expensive than just the cost of floor trusses." Biggin also likes how easily floor trusses install. Everything is precut and ready to go, so even Habitat volunteers who don't have framing experience can handle the installation without a glitch.

Biggin points out that proper planning and site management are the keys to capitalizing on the cost/time benefit of using trusses. Training installers and subs about the dos and don'ts goes a long way toward avoiding costly mistakes. An entire set of trusses can be installed and sheathed before anyone realizes that they've all been placed upside down; it has happened before. Paying attention to the manufacturer's instructions is important.

Open Webs Benefit Everyone

For shallower depths and shorter spans, floor trusses are more expensive than I-joists, but as Biggin points out, comparing the cost of I-joists to trusses without considering the big picture can be misleading. Builders who swear by trusses like them most for their open webs, which make mechanical contractors' work a lot easier.

"My subs are able to run plumbing, wiring, and any other utility without drilling holes," says Arkansas builder Gary Striegler. "They're through the job site faster, which moves the whole project along more quickly." For Striegler and other builders, saving time for subcontractors translates into saving money for the whole project.

HOW ENGINEERED JOISTS WORK

JOIST SPANS DEPEND ON THE STRENGTH and stiffness of the member and the amount of load it is required to carry. Deflection, or the degree to which a joist flexes under the design load, is commonly taken to be the main factor in the way a floor performs. Live loads (such as the weight of furniture and people) and dead loads (the weight of actual materials) are both used in calculating deflection.

The International Residential Code (IRC) limits floor-joist deflection to span/360 (where span is measured in inches) for live loads in living spaces (40 psf) and in sleeping areas (30 psf). The higher the denominator, the lower the deflection and, generally, the better the performance. Although this might generally be true, deflection isn't the only variable to consider in a floor's performance.

According to Tim Debelius, a spokesman for joist manufacturer iLevel (www.ilevel.com), a floor can have a deflection rating as high as span/720 and still feel bouncy, or as low as span/280 and feel firm. Joist depth plays an obvious role in performance. The deeper the joist, the stiffer the floor will likely be. Ceilings installed on the bottom of a floor system and other materials also affect performance. Ceilings, for example, help to brace the bottoms of joists, tying them all together and limiting their ability to shift left to right. Strongbacks, blocking, and flooring material matter, too. Factors including the floor system's weight (dead load) and the elasticity of these materials all contribute to a floor's performance.

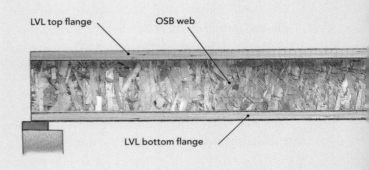

LVL top flange   OSB web

LVL bottom flange

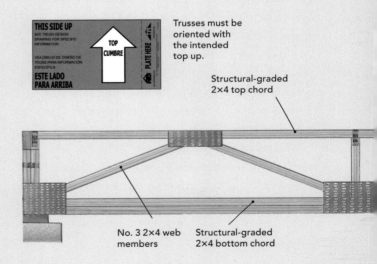

Trusses must be oriented with the intended top up.

Structural-graded 2×4 top chord

No. 3 2×4 web members

Structural-graded 2×4 bottom chord

Drill Holes with Caution

I-joist flanges and webbing each play a crucial role in the way load is distributed within the member. Modifying flanges in any way could result in structural failure. Drilling holes in the webbing outside of the manufacturer's recommendations could result in structural failure as well. See the manufacturer's installation information for specific guidance on hole sizing and spacing requirements. Below are examples of some requirements.

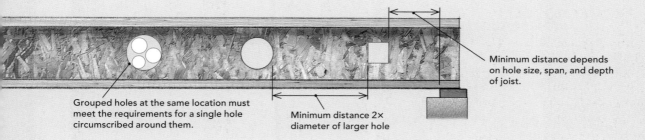

Minimum distance depends on hole size, span, and depth of joist.

Grouped holes at the same location must meet the requirements for a single hole circumscribed around them.

Minimum distance 2× diameter of larger hole

Don't Change a Thing

Because of the way trusses are engineered, every piece has a predetermined role in how it performs. Therefore, any modifications, however slight, affect the entire truss and could result in structural failure. Don't trim the ends, add point loads without an engineer's approval, or cut, notch, and drill through webs, plates, or chords. Also, look for installation guidance from tags like the ones shown here.

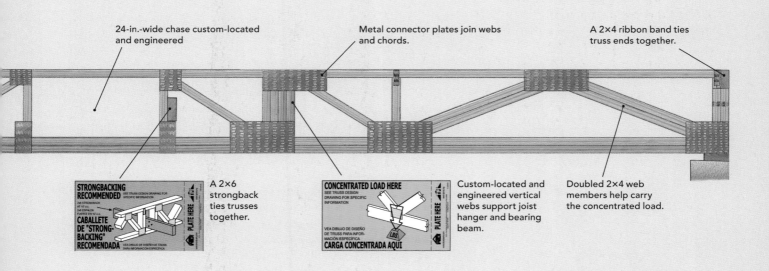

24-in.-wide chase custom-located and engineered

Metal connector plates join webs and chords.

A 2×4 ribbon band ties truss ends together.

A 2×6 strongback ties trusses together.

Custom-located and engineered vertical webs support joist hanger and bearing beam.

Doubled 2×4 web members help carry the concentrated load.

KEY INSTALLATION DETAILS

I-JOISTS

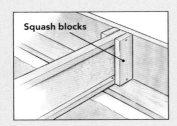

Squash blocks are required at bearing locations where concentrated loads from above must be transferred through the floor assembly. Blocks should be made of 2×4s or 2×6s, oriented vertically, and 1/16 in. longer than the depth of the joist.

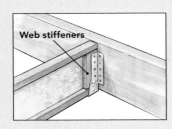

Web stiffeners are required to provide a nailing surface when certain types of hangers are used. The web stiffener should be made of either plywood or OSB sheathing or utility-grade spruce-pine-fir (SPF) lumber.

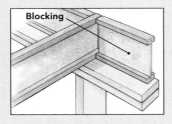

Blocking prevents joists from rolling, and transfers shear and vertical loads from above. Cutoffs from I-joists or structural rim boards can be used. Both APA–The Engineered Wood Association and the Wood I-Joist Manufacturers Association (WIJMA) recommend against using sawn lumber for blocking.

FLOOR TRUSSES

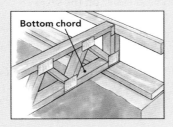

Bottom chord-bearing trusses sit on top of a mudsill, beam, or bearing wall. The ends are typically tied together with a 2× "ribbon board," which provides lateral stability during installation. Structural wall sheathing is often used for lateral stability. Special blocking panels may be required to transfer shear loads through the assembly.

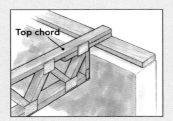

Top chord-bearing trusses are supported on the top of the mudsill, beam, or bearing wall by their top chords. They're used most often in applications that require the floor be close to grade level. They can't roll like bottom chord-bearing trusses, but they still require bracing before the subfloor is installed.

Michael Chandler uses floor trusses for the second floors of the houses he builds. Chandler's truss designer allocates a section of the truss for a utility chase. Using this design allows Chandler's subs to keep their work clean and neat. The greatest benefit for Chandler, though, is from an energy-efficiency standpoint. "It's important to have all HVAC duct-work within the building envelope. Trusses provide a place for ductwork in a conditioned environment, not an attic," Chandler says.

Dialing in Performance Is a Tricky Equation

It used to be that builders didn't have many op-tions or resources available for customizing floor performance. Sawn-lumber floor joists were typi-cally the same grade, size, and spacing required for the longest span within the system. With I-joists and trusses, builders can design and install floors that maximize performance based on each part of the house and the way it will be used.

Living spaces such as hallways don't require the same deflection rating as sleeping spaces, for ex-ample. No matter how spaces feel underfoot, build-ers want to ensure that floors don't make a sound, that tile doesn't crack, and that drywall doesn't pop. Joist sizing and strength can vary in different parts of a house.

Open-web floor trusses play a large role here. A builder can hand plans to the truss manufacturer with specs such as desired deflection, floor-material selections, and location of utility chases. In about a week, the builder gets a floor-framing plan that's fully customized and guaranteed to perform.

Custom-home builders such as Mike Guertin in Rhode Island like this level of customization and peace of mind. "I want to hand my plans over to someone and get back my layout and design op-tions," Guertin says. "I don't want to spend my time looking at span tables and punching numbers into a calculator in order to figure this out."

This level of customization is why it's difficult to say how much a floor truss will cost per lineal foot. Taft Ketchum from PDJ Components Inc. in Ches-ter, New York, points out that truss sourcing varies by region, too, which makes it hard to gauge pricing. "In our area, lumberyards act as the middleman, but in other parts of the country, builders deal directly with the truss plant," he says.

It's interesting to note that unlike with dimen-sional lumber and I-joists, deeper doesn't necessar-ily mean more expensive. Ketchum explains that depending on the design, a 12-in.-deep truss might be more expensive than a 16-in.-deep truss simply because of the amount of material used. A shallower truss may call for more webs, which means more wood, more labor, and ultimately, higher cost.

The bottom line is that you'll pay more for trusses and I-joists than for dimensional lumber. You'll also get a more stable, sometimes fully customized prod-uct that's faster to install and easier on the subs.

Managing Job-Site Mud

BY FERNANDO PAGÉS RUIZ

Drive past an average construction site (even a small residential addition) after a summer rain, and the street is usually coated with mud. Gooey, sticky, dirty stuff, the mud that runs off job sites and flows into storm sewers wreaks havoc on the quality of streams, rivers, and other waterways. But beyond the dire environmental consequences of job-site runoff, it's also rude to mire your neighbors in mud. Plus, there's the matter of steep fines.

It's the Law

Although most municipal ordinances include punitive measures for noncompliance, all the building officials I spoke with in my research focus on prevention through builder education and support rather than coercion. But they take the job seriously enough to prosecute those who don't cooperate. Penalties for job-site pollution range from stop-work notices to thousand-dollar-a-day fines and even criminal prosecution.

"It's a question of influencing the construction culture," says Terry Ullsperger, a watershed-management inspector for Lincoln, Nebraska, who describes himself as someone who "has been on both sides of the silt fence." Ullsperger likens the cultural conversion effort to the famous 1960s "Don't Be a Litterbug" campaign, which made it unthinkable to toss trash

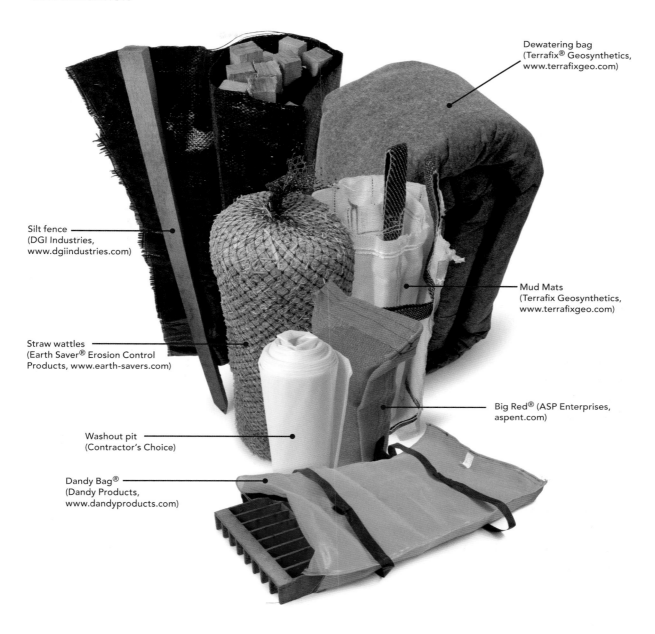

Dewatering bag
(Terrafix® Geosynthetics,
www.terrafixgeo.com)

Silt fence
(DGI Industries,
www.dgiindustries.com)

Straw wattles
(Earth Saver® Erosion Control
Products, www.earth-savers.com)

Mud Mats
(Terrafix Geosynthetics,
www.terrafixgeo.com)

Washout pit
(Contractor's Choice)

Big Red® (ASP Enterprises,
aspent.com)

Dandy Bag®
(Dandy Products,
www.dandyproducts.com)

from a car window. "Builders are slowly realizing a clean job site is just good building practice," says Ullsperger.

Similarly, Janice Lopitz of the Keep It Clean partnership in Boulder, Colorado, believes that those who would never wash a paintbrush in a stream bed may not realize they are doing the same thing when rinsing paint from their brushes at the curb. When you wash on the curb, the paint enters a storm-water inlet and heads straight to the nearest stream, lake, or river. "Whatever hits the street is as good as in the stream," says Lopitz.

Big builders have been on notice for several years. Federal standards require a storm-water pollution-prevention plan when construction extends over an acre of land. This plan explains in detail what you will do to keep pollutants, principally mud, from seeping into the storm-water system. It requires an

BREAK YOUR LOT INTO FOUR ZONES

To establish effective erosion and runoff controls on a job site, the first step involves walking the property to observe natural drainage patterns, potential hazards (such as a storm-water inlet in close proximity to the site), and the best areas for construction access and material handling. In essence, think of your job site as having four zones. Address each zone with the appropriate products and techniques.

ZONE 1: ESTABLISH A PERIMETER

The best method for controlling runoff is to preserve as much natural vegetation as possible. If the vegetation is removed or disturbed, you'll have to keep any eroding soil or washed-away sediments on the property through other means.

- **Silt fence** is made from woven polypropylene yarn designed to block sediment while letting water flow through it. Silt fence should be placed downslope of disturbed ground, and the stakes to hold the fence in place should be stocked on site.

- **Wattles,** also known as filter socks or fiber rolls, are essentially mulch sausages. The casing is a biodegradable mesh, and the stuffing is usually made of agricultural waste products. They are staked in place and work well when tiered on slopes.

ZONE 2: PROTECT STORM-WATER INLETS

The last line of defense comes at the storm-sewer inlet. A standard approach—and a wrong one—is to place a bale of hay in front of the inlet. Bales break down quickly and dam water, or divert it someplace else. The real goal is to filter sediment out of the water entering the inlet.

- **Dandy Bag** by Dandy Products is a filter designed for use with flat grates and mountable curbs. The Dandy Bag is made of high-strength filter fabric. The inlet grate is placed in the bag before being placed back in its location.

- **Big Red** by ASP Enterprises is a highly porous filter sock that simply lies in front of an open throat-style inlet to prevent sludge from entering the storm-water line. The filter sock can be positioned to allow clean water to flow over it and/or through it.

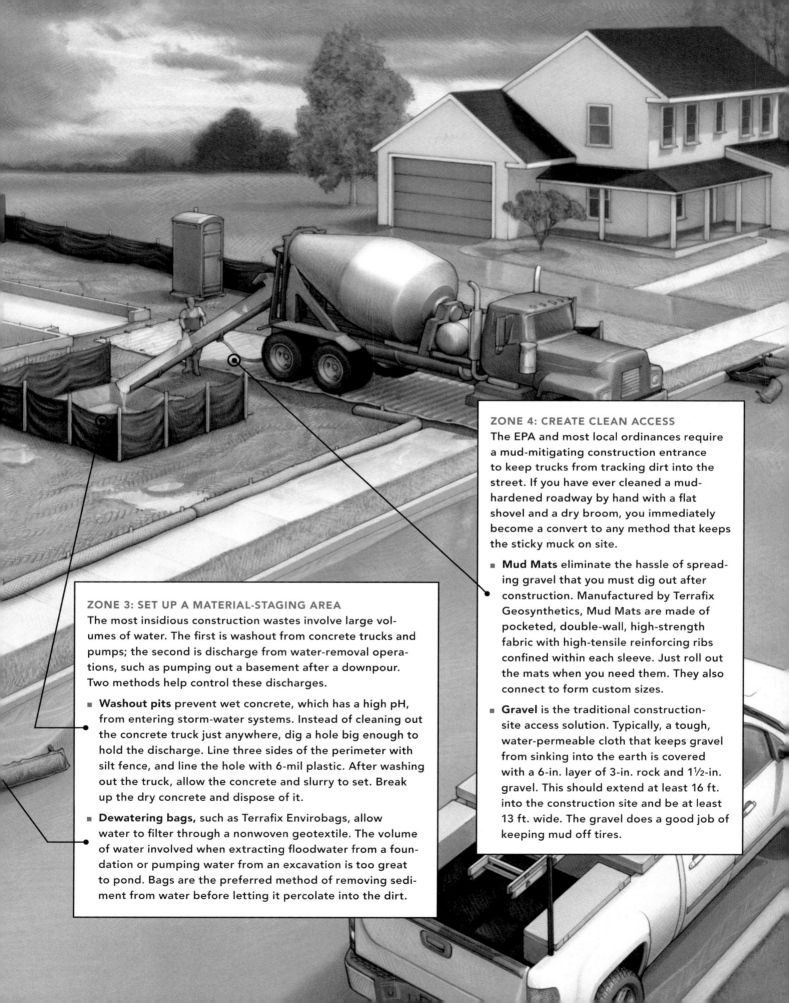

ZONE 4: CREATE CLEAN ACCESS

The EPA and most local ordinances require a mud-mitigating construction entrance to keep trucks from tracking dirt into the street. If you have ever cleaned a mud-hardened roadway by hand with a flat shovel and a dry broom, you immediately become a convert to any method that keeps the sticky muck on site.

- **Mud Mats** eliminate the hassle of spreading gravel that you must dig out after construction. Manufactured by Terrafix Geosynthetics, Mud Mats are made of pocketed, double-wall, high-strength fabric with high-tensile reinforcing ribs confined within each sleeve. Just roll out the mats when you need them. They also connect to form custom sizes.

- **Gravel** is the traditional construction-site access solution. Typically, a tough, water-permeable cloth that keeps gravel from sinking into the earth is covered with a 6-in. layer of 3-in. rock and 1½-in. gravel. This should extend at least 16 ft. into the construction site and be at least 13 ft. wide. The gravel does a good job of keeping mud off tires.

ZONE 3: SET UP A MATERIAL-STAGING AREA

The most insidious construction wastes involve large volumes of water. The first is washout from concrete trucks and pumps; the second is discharge from water-removal operations, such as pumping out a basement after a downpour. Two methods help control these discharges.

- **Washout pits** prevent wet concrete, which has a high pH, from entering storm-water systems. Instead of cleaning out the concrete truck just anywhere, dig a hole big enough to hold the discharge. Line three sides of the perimeter with silt fence, and line the hole with 6-mil plastic. After washing out the truck, allow the concrete and slurry to set. Break up the dry concrete and dispose of it.

- **Dewatering bags,** such as Terrafix Envirobags, allow water to filter through a nonwoven geotextile. The volume of water involved when extracting floodwater from a foundation or pumping water from an excavation is too great to pond. Bags are the preferred method of removing sediment from water before letting it percolate into the dirt.

engineer's stamp and inspection of mitigation methods every two weeks and after every storm. The plan also requires a living, breathing individual (not just a business entity) to become responsible and liable for the methods used, their maintenance, and their effectiveness.

Small sites are governed locally and increasingly require a permit with a simple plot plan illustrating the lot's drainage pattern and the methods you will use to mitigate erosion, runoff, and pollutants.

Good Housekeeping

Some builders have embraced the new job-site management practices and have discovered an unexpected benefit. "Customers notice a clean job site and assume our construction is as good as our housekeeping," says Sean Smetter of Smetter Custom Homes in Lincoln, Nebraska. Smetter attributes at least part of his success in a tough economy to customers seeing his tidy job site as evidence of the quality consciousness they were looking for in a builder.

But maintaining white-glove standards on a job site requires constant vigilance. You have to check perimeter erosion-control systems at least every two weeks and after every storm. You have to spade accumulated silt off the storm-sewer inlet barrier. You have to restake silt fences and reposition wattles. And after your favorite subcontractor drives off into the sunset, leaving a trail of mud behind his pickup, guess what? It's your responsibility to make sure the street has been swept clean before sundown.

Consider Subbing It Out

In response to the ratcheting up of federal and municipal job-site pollution-control requirements, a new class of geotech subcontractor has evolved. Mitigation experts can take the headache of designing, installing, and maintaining storm-water management off your to-do list.

Outfits like Soil-Tek® (www.soil-tek.com) in the Midwest, Down to Earth Compliance (DTEC; www.trustdtec.com) in the Mountain States, and Acacia Erosion Control (www.acaciaerosioncontrol.com) on the West Coast have the necessary certifications and equipment to make the job easy.

These subs not only have the tractors to knife in silt fence, but also offer the latest in geotech products and biofilter technology, which is used to re-establish erosion and sediment control after construction.

Temporary Power on the Job Site

BY JOSEPH FRATELLO

Virtually everything we use on a job site, from compressors and battery chargers to fax machines and computers, requires electricity. Until a project has passed its final electrical inspection, all but the smallest construction sites require temporary power, whether it's provided by a portable generator or through temporary electrical service.

Over the past few years the shrinking scale and budgets of many projects have underscored how overlooked the issue of temporary power has become. Recently I started a fairly large project that was expected to last just over a year. The builder refused to pay for a temporary service, so I learned just how miserable and dangerous it was not to have sufficient power available at the job site.

What Exactly Is Temporary Power?

The definition of temporary power depends on whom you ask. The National Electrical Code lists its definition in article 590 of publication NFPA 72: "Installations may be considered temporary as long as you are in the process of construction, remodeling, maintenance, repair, demolition of buildings or similar activities; the temporary power will be permitted for the length of the time needed to complete the project." Your local building department

THE SECOND-BEST OPTION: GENERATORS

PORTABLE GENERATORS ARE A convenient way to power up a job site. The better models provide enough power to supply a compressor and a couple of saws. However, they do have limitations.

- Limited output (typically 3,500w to 6,500w)
- Recurring cost of maintenance and fuel
- Noise

SURGE PROTECTORS AND GENERATORS GO TOGETHER. When running on generator power, use an uninterruptible power supply (UPS) strip for battery chargers, phones, and other electronics. The generator will run more smoothly, and the UPS's battery backup keeps electronics powered for up to an hour in the event of a generator failure.

EXTENSION-CORD BASICS

- Keep the cord length as short as possible.
- Use 12/3 wire. Anything smaller creates resistance, which in turn can wear out tools.
- Replace cords that have broken plugs, tears in the jacket, or exposed conductors.

probably has its own interpretation of "temporary," and of course, there are OSHA guidelines. Although all may be similar, I follow the NEC guidelines.

A common mistake I encounter on job sites is the assumption that temporary power doesn't have to adhere to electrical codes because it's temporary. The hazards of electricity don't change just because it is temporary, however. I was recently on a job where another electrical contractor did a horrendous job installing the temporary power, and as a result, a fire started. After investigating the cause, I discovered the contractor had installed the wrong size breaker and failed to secure and terminate the wires properly. When installing temporary power, you must still adhere to NEC rules for wire sizing, grounding, and wiring methods.

Is a Generator a Good Idea?

When starting a project, you need to identify the source of your power. Typically, your choices are either a generator or the utility company. (Alternative-energy sources are gaining market share, but they are not widely used yet.) Many contractors on small to medium-size jobs use portable generators to provide temporary power, which is a convenient way to get started at a job site. Most generators come with multiple receptacles that can handle a few tools. (I own a few generators for a variety of tasks. If you don't already own a generator, I highly recommend buying a Honda®. Yes, they are more expensive than most, but they are well worth it. After 20 years in the construction business, I have never seen one break down, and I have never brought one in for service.)

Even the best portable generators have drawbacks, though, so I try to avoid them. With a generator, you are limited by its output. Most generators that can be moved by one or two people can output only 3,500w to 6,500w, which is barely sufficient to power a compressor and a circular saw at the same time. Straining your tools on a generator can lead to premature failure. If you plan on running a tablesaw, a compressor, a miter-saw station, and some temporary lights, most portable generators are not up to the job.

Generators also require maintenance and constant refueling. My generator can use 3 gal. of gas in a day, which costs more than $75* a week. That doesn't include the labor for refueling and changing oil.

Generator-sourced power can create other problems, too. If you're working in a basement and the generator runs out of gas, you can be stuck in the dark trying to find your way out. Many generators also tend to make "dirty," or unstable, power, which can affect battery chargers and other sensitive electronics. Conversely, battery chargers can affect the generator. If you've ever noticed that your generator runs erratically when you plug in your battery charger, it's because the charger's constant off-on pulse messes with the generator. To fix this and to protect other sensitive electronics, get a battery-equipped uninterruptible power supply (UPS) surge protector. These devices typically cost less than $100, which is a small price to pay to protect your phone, your chargers, and your laptops.

Another drawback of generators is that they tend to be noisy. This is especially problematic where I work. Our workday starts at 6:30 a.m., but our township will not allow contractors to start generators until 8 a.m. Some localities also have ordinances that limit a generator's decibel level, no matter what time of day.

Temporary Service Sometimes Makes the Most Sense

After a month of rolling up multiple 100-ft. cords and refueling my generator, I had enough and absorbed the cost of installing a temporary service. I quickly realized that it would pay for itself after a month. If you are on a job where the builder doesn't want to cover the cost of a temporary service, you might consider asking the other trades to chip in. Typically, a temporary service costs from $600 to $1,200 to install. I save the poles and equipment and move them from job to job, and typically charge my builders the permit fees and labor to dig the pole into the ground. It may seem like overkill, but I install 100 amps for most of my temporary services.

Because I have such a wide variety of projects, it's easier just to make them all 100 amp and to reuse the equipment.

Safe Wiring Is the Next Step

Once the decision has been made to go with temporary service, you have to make sure to run wiring safely throughout the project and to avoid the numerous hazards created by temporary wiring installations. I have seen two small fires as a result of poorly installed temporary wiring. Electrocution is certainly a possibility. In my experience, though, the greatest hazard resulting from poorly installed temporary wiring is tripping. I have seen dozens of people (including myself on occasion) trip and fall, and a few have ended up at the hospital with a broken wrist or a sprained ankle.

Here's how I like to set up most of my larger jobs. I start by providing two dedicated 20-amp GFCI receptacles outside with deep in-use covers. All temporary circuits must be protected with GFCI breakers or outlets. For receptacles that see infrequent use, choose those rated as weather-resistant GFCIs. Receptacles that get more use will not last as long, so it's more economical to use non-GFCI receptacles tied to GFCI breakers. To keep the receptacles dry, I prefer the deep in-use covers because they allow enough room for heavy-gauge extension cords to bend without kinking the ends.

When the framing is completed and the project is weathertight, I ask the carpenters if they need any specialty outlets for stationary tools. I start by providing a minimum of two 20-amp GFCI receptacles on each floor, one for power tools and one for temporary lighting. On smaller jobs, I run the 12/2 NM home runs for the bathroom GFCI receptacles as soon as possible, and connect them to a temporary panel in the basement. If you already have power, energize them first.

On bigger jobs, I install a subpanel close to the building's main entrance to feed the outlets on the

(Continued on p. 40)

STAGE ONE: SET UP A PANEL OUTSIDE

DURING THE EARLY STAGES OF CONSTRUCTION, a temporary panel is often the source for job-site electricity. Commonly mounted on a separate pole that's buried about 4 ft. into the ground, the panel is connected to the street and should have 100-amp service. Outlets can be mounted adjacent to the panel or remotely.

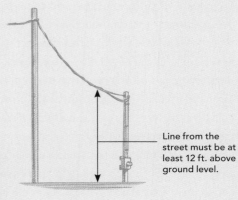

Line from the street must be at least 12 ft. above ground level.

PROTECT CIRCUITS OUTSIDE. Installed in temporary panels, GFCI breakers can provide the required protection for non-GFCI receptacles on that circuit. A GFCI breaker should also be used when protecting a 240v receptacle.

USE THE PROPER RECEPTACLE, AND COVER IT. Any receptacle used outside must be either a weather-resistant GFCI-protected model or connected to a protected circuit, and by code it must be installed in a covered box. Deep covered boxes keep the outlets dry and also provide strain relief so that extension-cord ends won't become kinked. It's always a good idea to install both 120v and 240v receptacles.

FOR CODE AND SAFETY. The panel must be connected to a ground rod, but check with local authorities because some jurisdictions require two grounds per panel.

REMOTE OPTION. Temporary receptacles also can be installed away from the panel in plastic posts. Power is supplied by 12/3 UF (underground feeder) cable that can be run below grade up to 250 ft. from the panel.

STAGE TWO: MOVE THE PANEL INSIDE

AFTER THE STRUCTURE IS CLOSED IN, IT OFTEN makes sense to move the panel inside. On larger jobs, the most convenient place to locate a temporary panel is by the main entrance. GFCI-protected receptacles can be installed in deep boxes at the panel or in more central locations.

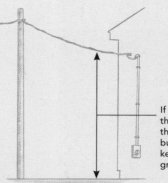

If the line from the street to the house is not buried, it must be kept 15 ft. above ground level.

KEEP THE LIGHTS ON A SEPA-RATE CIRCUIT. It's a good idea to provide a separate circuit and switch for the temporary lights. If someone trips a breaker with a tool, they won't be left in the dark. The switch is a convenient way to control the lighting.

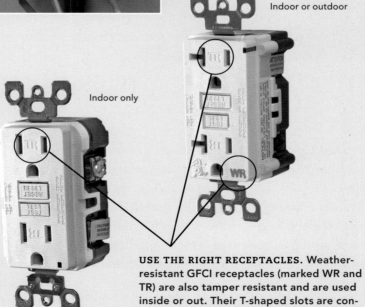

Indoor or outdoor

Indoor only

USE THE RIGHT RECEPTACLES. Weather-resistant GFCI receptacles (marked WR and TR) are also tamper resistant and are used inside or out. Their T-shaped slots are configured for 20-amp plugs. Tamper-resistant-only models (marked TR) are meant to be used indoors.

THE MOST IMPORTANT LESSONS FOR TEMPORARY LIGHTING

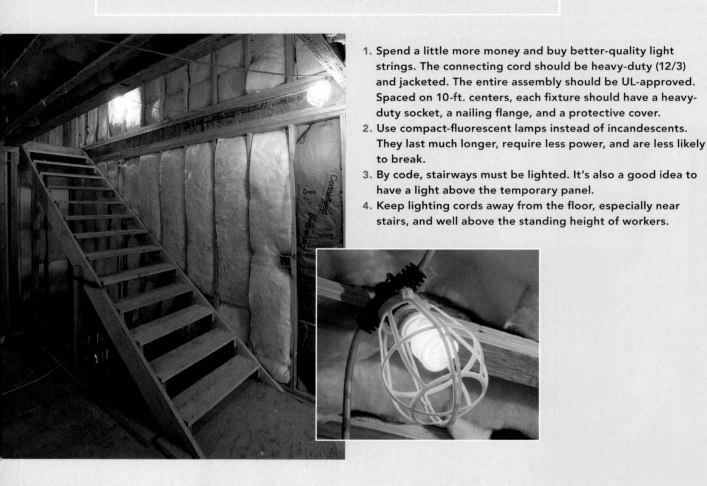

1. Spend a little more money and buy better-quality light strings. The connecting cord should be heavy-duty (12/3) and jacketed. The entire assembly should be UL-approved. Spaced on 10-ft. centers, each fixture should have a heavy-duty socket, a nailing flange, and a protective cover.
2. Use compact-fluorescent lamps instead of incandescents. They last much longer, require less power, and are less likely to break.
3. By code, stairways must be lighted. It's also a good idea to have a light above the temporary panel.
4. Keep lighting cords away from the floor, especially near stairs, and well above the standing height of workers.

first floor. I prefer that location because it's far more convenient to walk a few feet rather than go outside to reset a tripped breaker. I also can use less expensive indoor-rated wire to go from the subpanel to the other outlets. Plus, at the end of the day I like being able to turn off the power and lights from the breakers in the subpanel and then lock the door behind me.

Consider Data Lines, Too

When running temporary power, it's also a good idea to install phone or cable lines to provide Internet access. Internet access not only provides the ability to transfer data such as blueprints and invoices, but it also can provide the ability to monitor the site with IP security cameras. Many of the projects we work on require temporary motion and fire alarms to maintain insurance qualifications. If your insurance provider doesn't require either, ask if using them will decrease your rates. The difference in cost may be well worth it. The ability to transfer data should be considered part of any significant project.

*Prices are from 2012.

General Framing

10 Rules for Framing

BY LARRY HAUN

I t was a coincidence that another contractor and I began framing houses next door to one another on the same day. But by the time his house was framed, mine was shingled, wired, and plumbed. It was no coincidence that the other contractor ran out of money and had to turn the unfinished house over to the lending company, whereas I sold mine for a profit.

Both houses were structurally sound, plumb, level, and square, but every 2×4 in the other house was cut to perfection. Every joint looked like finish carpentry. The other contractor was building furniture, and I was framing a house.

Unlike finish carpentry, framing doesn't have to look perfect or satisfy your desire to fit together two pieces of wood precisely. Whether you're building a house, an addition, or a simple wall, the goals when framing are strength, efficiency, and accuracy. Following the building codes and the blueprints should take care of the strength; efficiency and accuracy are trickier. But during 50 years of framing houses, I've come up with the following rules to help me do good work quickly and with a minimum of effort.

1. Don't Move Materials Any More Than You Have To

Hauling lumber from place to place is time-consuming and hard on your body. Make it easier on yourself every chance you get, and start by having the folks at the lumberyard do their part. Make sure lumber arrives on the truck stacked in the order it will be used. You don't want to move hundreds of wall studs to get to your plate stock, for instance. And floor joists go on top of floor sheathing, not the other way around.

When it's time for the delivery, unload the building materials as close as possible to where they will be used. Often lumber can be delivered on a boom truck, so stacks of lumber can be placed right up on the deck or on a simple structure built flush alongside the deck.

Once the material is delivered, don't move it any more than you need to. Cut studs, plywood, and anything else you can right on the stack. If you do have to move wood, plan so that you have to move it only once.

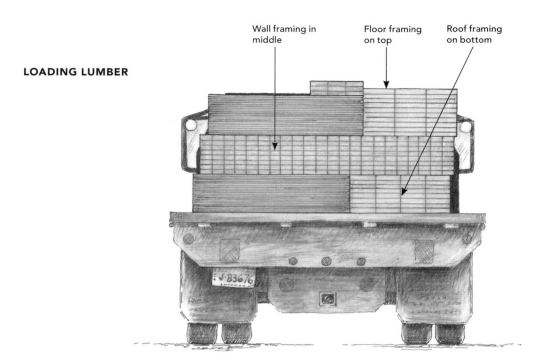

Wall framing in middle

Floor framing on top

Roof framing on bottom

2. Build a House, Not Furniture

In other words, know your tolerances. Rafters don't have to fit like the parts of a cabinet. Nothing in frame carpentry is perfect, so the question is this: What's acceptable?

You do need to get started right, and that means the mudsills. Whether they're going on a foundation or on a slab, they need to be level, straight, parallel, and square. But there's no harm done if they're cut ¼ in. short (see the drawing at left on p. 44). A rim joist, on the other hand, needs to be cut to the right length (within ¹⁄₁₆ in.) before being nailed to the mudsill.

When it comes to wall framing, the bottom plate also can be ¼ in. or so short, but the top plate needs to be cut to exact length (again within ¹⁄₁₆ in.) because it establishes the building's dimension at the top of the walls. But the plate that sits on top of that, the cap or double plate, should be cut ¼ in. short so that intersecting walls tie together easily.

Once you've raised the walls, how plumb or straight is good enough? In my opinion, ¼ in. out of plumb in 8 ft. is acceptable, and a ¼-in. bow in a 50-ft. wall won't cause harm to the structure or

Cut, Don't Move, the Stack
Cut 2×4s right on the stack.

problems for subcontractors. Take special care by framing as accurately as possible in the kitchen and in the bathrooms. These rooms require more attention, partly because of their tighter tolerances, but also because the work of so many trades comes together here.

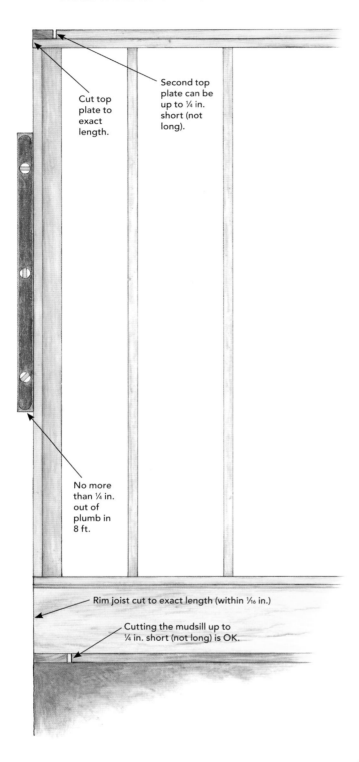

Cut top plate to exact length.

Second top plate can be up to ¼ in. short (not long).

No more than ¼ in. out of plumb in 8 ft.

Rim joist cut to exact length (within ¹⁄₁₆ in.)

Cutting the mudsill up to ¼ in. short (not long) is OK.

3. Use Your Best Lumber Where It Counts

These days, if you cull every bowed or crooked stud, you may need to own a lumber mill to get enough wood to frame a house. How do you make the most of the lumber that you get?

Use the straightest stock where it's absolutely necessary: where it's going to make problems for you later on if it's not straight (see the drawing below). Walls, especially in baths and kitchens, need to be straight. It's not easy to install cabinets or tile on a wall that bows in and out. And straight stock is necessary at corners and rough openings for doors.

The two top wall plates need to be straight as well, but the bottom plate doesn't. You can bend it right to the chalkline and nail it home. If you save your straight stock for the top plates, you'll have an easy time aligning the walls. And every project needs lots of short stock for blocking; take your bowed material and cut it into cripples, headers, and blocks.

4. Work in a Logical Order

Establish an efficient routine for each phase of work, do it the same way every time, and tackle each phase in its logical order. In the long run, having standard procedures will save time and minimize mistakes. Let's take wall framing as an example.

First I snap all of the layout lines on the floor; then I cut the top and bottom plates and tack all of them in place on the lines. Next I lay out the plates, detailing the location of every window, door, stud, and intersecting wall.

Pick the Right Stock for the Right Place

Straight lumber (top) is important for many locations, but some places, like the bottom wall plate, don't need perfectly straight stock.

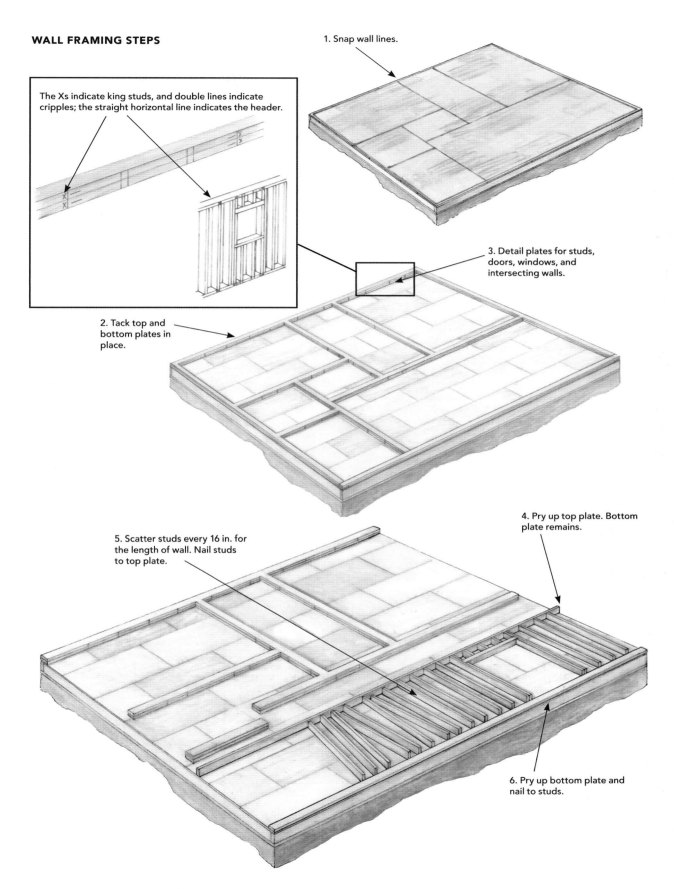

WALL FRAMING STEPS

The Xs indicate king studs, and double lines indicate cripples; the straight horizontal line indicates the header.

1. Snap wall lines.

2. Tack top and bottom plates in place.

3. Detail plates for studs, doors, windows, and intersecting walls.

4. Pry up top plate. Bottom plate remains.

5. Scatter studs every 16 in. for the length of wall. Nail studs to top plate.

6. Pry up bottom plate and nail to studs.

I pry up the top plate and move it about 8 ft. away from the bottom plate, which I leave tacked to the deck. I scatter studs every 16 in. for the length of the wall. I nail the top plate to the studs and keep the bottom of the studs snug against the bottom plate. This helps to keep the wall square, straight, and in position to be raised. I try to establish a rhythm and work consistently from one end to the other. Once the top plate is completely nailed, I pry up the bottom plate and repeat the process on the bottom.

It's worth saying that I didn't just make up these steps; they evolved over time. Recognizing inefficiency is an important part of framing.

5. Keep the Other Trades in Mind

If you want to waste time and money when framing, don't think about the electrical work, the plumbing, the heat ducts, the drywall, or the finish carpentry. Whether you do them yourself or hire subcontractors, these trades come next. And unless you're working with them in mind every step of the way, your framing can be in the way.

For example, when you nail on the double top plate, keep the nails located over the studs. This tip leaves the area between the studs free for the electrician or plumber to drill holes without hitting your nails.

PLAN FOR ALL THE PIECES, NOT JUST FRAMING

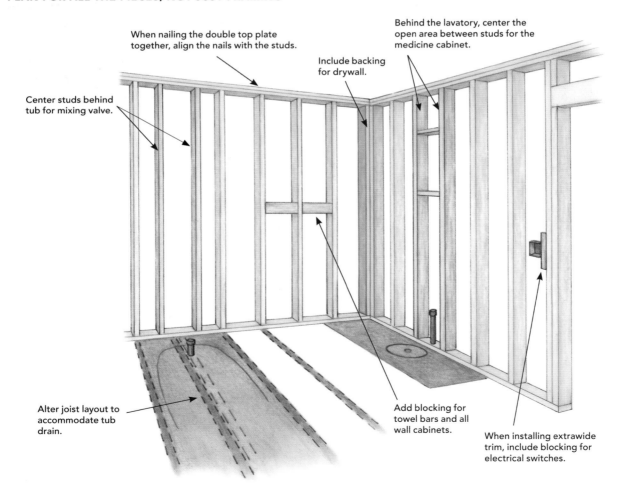

When nailing the double top plate together, align the nails with the studs.

Behind the lavatory, center the open area between studs for the medicine cabinet.

Include backing for drywall.

Center studs behind tub for mixing valve.

Alter joist layout to accommodate tub drain.

Add blocking for towel bars and all wall cabinets.

When installing extrawide trim, include blocking for electrical switches.

6. Don't Measure Unless You Have To

The best way to save time when you're framing a house is by keeping your tape measure, your pencil, and your square in your nail pouch as much as possible. I have to use a tape measure to lay out the wall lines accurately on the deck, but after that, I cut all of the wall plates to length by cutting to the snapped wall lines. I position the plate on the line, eyeball it, and then make the cuts at the intersecting chalkline.

Another time-saver is to make square crosscuts on 2×4s or 2×6s without using a square. Experience has shown me that with a little practice, anyone can make these square cuts by aligning the leading edge of the saw's base, which is perpendicular to the blade, with the far side of the lumber before making the cut.

Trimming ¼ in. from a board's length shouldn't require measuring. Ripping (lengthwise cuts) longer pieces also can be done by eye if you use the edge of the saw's base as a guide. Train your eye. It'll save time cutting, and as you develop you'll also be able to straighten walls as easily by eye as with a string.

7. Finish One Task before Going on to the Next

My first framing job was with a crew that would lay out, frame, and raise one wall at a time before moving on to the next. Sometimes they would even straighten and brace the one wall before proceeding. We wasted a lot of time constantly switching gears.

If you're installing joists, roll them all into place and nail them before sheathing the floor. Snap all layout lines on the floor before cutting any wall plates, then cut every wall plate in the house before framing. If you're cutting studs or headers and cripples, make a cut list for the entire project and cut them all at once. Tie all the intersecting walls together before starting to straighten and brace the walls.

Finishing before moving on is just as important when it comes to nailing and blocking. You might be tempted to skip these small jobs and do them later, but don't. Close out each part of the job as well as you can before moving on to the next. Working in this way helps to maintain momentum, and it prevents tasks from being forgotten or overlooked.

Use Your Eye, Not Your Tape Measure

With practice, you can make square cuts by aligning the front edge of the saw's base with the far edge of the board.

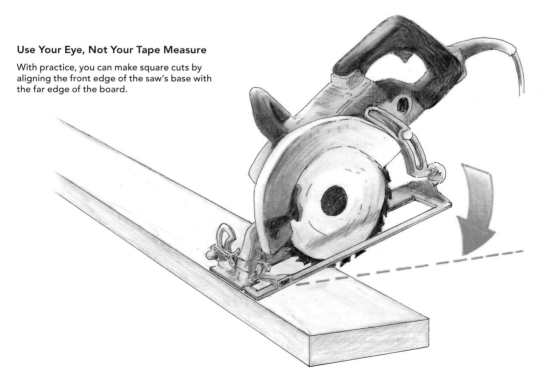

8. Cut Multiples Whenever Possible

You don't need a mathematician to know that it takes less time to cut two boards at once than it does to cut each one individually.

If you have a stack of studs that all need to be cut to the same length, align one end of the top row, snap a chalkline all the way across, and cut the studs to length right on the pile. Or you can spread them out on the floor, shoving one end against the floor plate, snap a chalkline, and cut them all at once.

Joists can be cut to length in a similar way by spreading them out across the foundation and shoving one end up against the rim joist on the far side. Mark them to length, snap a line, and cut the joists all at once.

Also, don't forget to make repetitive cuts with a radial-arm or chop/miter saw outfitted with a stop block, which is more accurate and faster than measuring and marking one board at a time.

9. Don't Climb a Ladder Unless You Have To

I don't use a ladder much on a framing job except to get to the second floor before stairs are built. Walls can be sheathed and nailed while they're lying flat on the deck. Waiting until the walls are raised to nail on plywood sheathing means you have to work from a ladder or a scaffold. Both are time-consuming.

With a little foresight, you can do the rafter layout on a double top plate while it's still on the floor. Otherwise, you'll have to move the ladder around the job or climb on the walls to mark the top plate.

10. Know the Building Code

Building codes exist to create safe structures. Because building inspectors are not capable of monitoring all parts of every project, it's your responsibility to know the building code and to build to it.

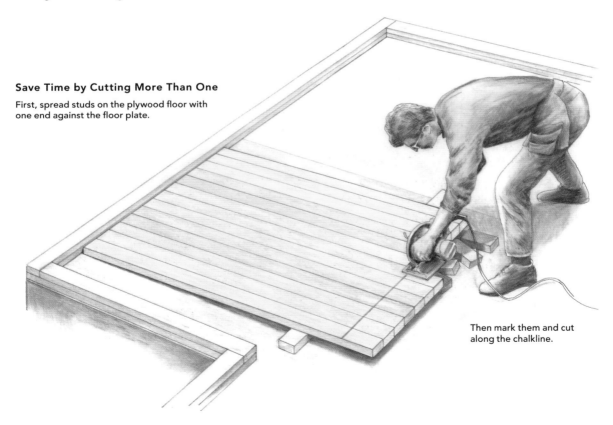

Save Time by Cutting More Than One

First, spread studs on the plywood floor with one end against the floor plate.

Then mark them and cut along the chalkline.

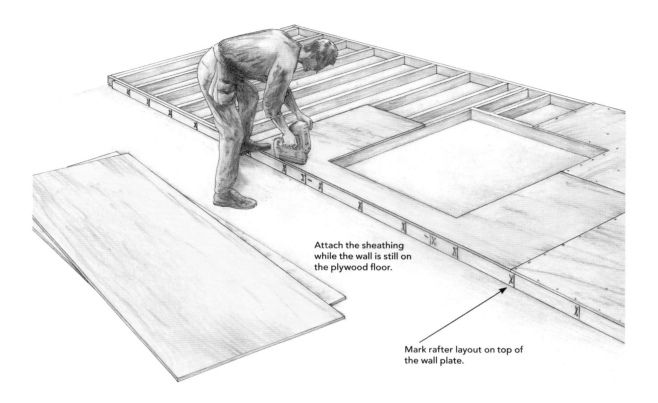

Attach the sheathing
while the wall is still on
the plywood floor.

Mark rafter layout on top of
the wall plate.

For instance, the code actually specifies how to nail a stud to a wall plate. You need two 16d nails if you're nailing through a plate into the end of the stud, or four 8d nails if you're toenailing. When you nail plywood or oriented strand board (OSB) roof sheathing, you need a nail every 6 in. along the edge of the sheathing and every 12 in. elsewhere. And if you're using a nail gun, be careful not to overdrive the nails in the sheathing.

A final word: If special situations arise, consult the building inspector. He or she is your ally, not your enemy. Get to know the building code for your area. Get your own copy of the *IRC (International Residential Code)* and build well, but build efficiently, with the understanding that perfection isn't what is required.

WORK SAFELY WHATEVER THE RULE

WORKING SAFELY should be at the top of your priority list. Safety glasses, hearing protection, and a dust mask should be the norm, as should attention around coworkers or dangerous debris.

Safety devices and good intentions, however, won't help if your mind isn't on the work. Pay attention, approach the work with a clear head, listen to that inner voice that says, "This is too dangerous," and be extra careful toward the end of the day.

Anchoring Wood to a Steel I-Beam

BY JOHN SPIER

Modern floor plans are trending toward wide-open spaces. Despite advances in engineered-wood beams, there are times when something stronger is needed. Many carpenters shy away from steel because fastening lumber to steel can be tricky. Cutting a steel beam on site is even trickier. Sometimes, though, a steel I-beam is the best choice. Structural steel costs less than comparable LVLs, is strong, and is available from local suppliers. If you order it to the right size with fastener holes punched, your only challenge will be attaching the lumber.

Steel Has a Few Limitations

Although a piece of steel carries a larger load over a longer span with less depth than any other building material, steel has some disadvantages. First, it's very heavy. You need to make sure you can get it to where it needs to go, either with humans or with a machine. Second, you won't find steel span charts in a codebook; steel usually needs to be sized by an engineer. Steel should be protected from moisture to prevent rust and deterioration. Also, a steel beam will fail much more quickly and catastrophically than an equivalent wood beam in a fire.

Top-Mount Joist Hangers Are the Most Practical

In the simplest usage, an I-beam rests in pockets cast in foundation walls, with floor joists on top of it. More often, though, wood is bolted to the web; then joists or rafters are attached to the wood with standard joist hangers. The drilling and bolting required by this method are so impractical that it's worth changing before construction. Another attachment method is to weld top-mount hangers to the I-beam. However, this option is rare, not because it doesn't work but because most framing crews don't have a welder on site.

I think the best method is to bolt 2× lumber to the top flange of the I-beam, then nail top-mount joist hangers to the lumber. I use ¼-in. by 1½-in. lag bolts through the bottom of the top flange. Because top-mount joist hangers won't resist twisting as well as face-mount hangers, I use strapping below or blocking between the joists. If uplift resistance is needed, I tack the lumber in place with a powder-actuated nailing tool, then through-bolt after the plywood subfloor is down, recessing the bolt heads into the plywood.

THE FASTEST WAY TO MARRY WOOD TO STEEL

Bolt a 2×6 or 2×8 to the top flange of a steel I-beam and use top-mount joist hangers to support the floor framing. Fastening 1½-in. lag bolts through the bottom of the top flange is far faster than countersinking the heads of through-bolts from above.

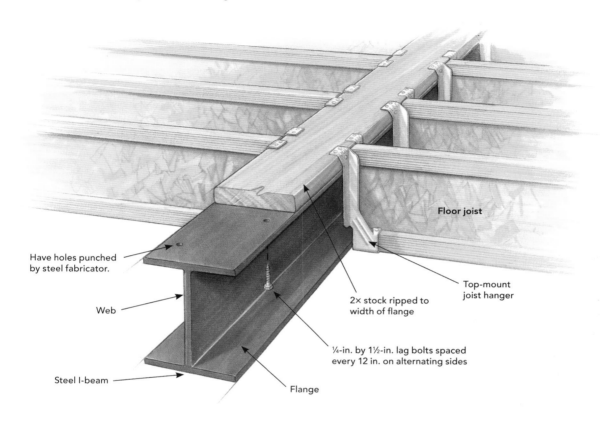

Have holes punched by steel fabricator.

Web

Steel I-beam

Flange

2× stock ripped to width of flange

¼-in. by 1½-in. lag bolts spaced every 12 in. on alternating sides

Floor joist

Top-mount joist hanger

A WASTE OF TIME

Packing the web is a last resort. Unless there are compelling reasons (such as headroom issues) to attach wood this way, I avoid it. This is a very strong method, but it's too time-consuming to make it practical for everyday use.

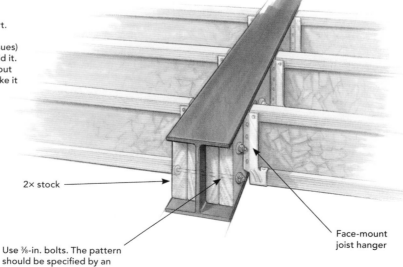

2× stock

Use ⅜-in. bolts. The pattern should be specified by an engineer.

Face-mount joist hanger

I-BEAM OPTIONS

MOST I-BEAMS ON JOB SITES ARE W-shapes, or wide flange. Another common beam is S-shape, or standard, which has narrower flanges and a thicker web. S-shapes are generally taller than comparable W-shape beams. Narrower S-shape beams are better suited for recessing into wall framing, whereas W-shape beams are great where headroom is an issue.

W-shape

S-shape

The steel I-beams most often seen on job sites are called W (wide) and S (standard) shapes, depending on the width of the flange. Steel beams are designated by shape, depth, and weight. For example, a W8×35 beam is a W-shape about 8 in. deep and weighing 35 lb. per lin. ft.

Because steel is hard to cut and drill on site, your supplier needs the exact length along with sketches showing all hole locations and sizes. If you have to cut a beam, an acetylene torch is the easiest method, but metal-cutting blades in circular saws and reciprocating saws work, too.

Steel Needs Protection from the Weather

To prevent corrosion, most suppliers spray steel with a primer. It's worthwhile to make sure primer is applied because it keeps the steel clean and rust-free while you're working with it. Occasionally, galvanizing is specified for steel components. If it is, make sure all the cutting and drilling are done first, and remember that the holes need to be oversize by $\frac{1}{8}$ in., or the bolts won't fit through after galvanizing is done.

Really big beams are best set in place with a crane, but lesser beams often can be placed with a lumber-delivery boom truck. A backhoe or excavator works if it can get close enough. I've found that beams as heavy as 600 lb. or so can be set safely with human power (four to six people), as long as they're not too high up. Use levers, rollers, winches, platforms, and plenty of caution. Finally, remember that wood shrinks and swells, but steel won't budge.

Wall Framing Basics

BY ROB MUNACH

When constructed correctly, a house's frame resists a variety of loads, such as wind and snow. When built poorly, the frame will fail. When overbuilt, it can lead to energy losses through thermal bridging.

To be a successful framer, you should have a comprehensive understanding of wall assemblies and employ advanced-framing techniques—including those illustrated here—when appropriate. Keep in mind that although advanced framing reduces energy losses, resource consumption, and construc-tion costs by using significantly less lumber than the typically framed house, it lacks the redundancy to stand up to extreme loads like a falling tree. So use advanced-framing techniques sensibly.

By looking at the components of a wall assembly and the role each component plays within a wall, you can begin to understand how a wall functions as a system. That's critical information, whether you're following standard building practices or advanced techniques. Here's how it works.

WALL FRAMING

HEADERS transfer wind-uplift loads and dead loads around openings in a wall assembly. They also transfer wind loads that blow on the face of the wall into the king studs. If exterior walls are not load bearing, headers can be constructed similar to the sill using a stud on the flat. In this case, a header has to resist wind loads blowing only on the face of the wall.

Installed on the flat, the **SILL PLATE** supports the window and transfers wind loads blowing on the face of the wall into the king studs. In a lot of cases, only a single sill plate is needed, not a double. Multiple sill plates are sometimes needed in wide openings.

Spaced 24 in. on center, 2×6 **STUDS** carry the downward load of the rafters, joists, or top plates above them. The studs in a wall also resist wind blowing against the wall and wind-uplift loads in walls that are not fully sheathed with plywood or OSB. This wall assembly yields more space for insulation than a wall with 2×4s spaced 16 in. on center.

CRIPPLES are short 2× blocks above and below rough openings. Those above the header transfer wind and gravity loads from the top plate to the header. They also give you something to nail sheathing to. The cripples below a sill serve mostly as nailers for the sheathing, so the number that you use can be reduced to only one in most instances.

FLOOR JOISTS resist live and dead loads. When joined with the rim joist, they create a system that resists racking forces on the house.

The **TOP PLATES** carry the downward load from the roof rafters or floor joists above. They also act as a "shear collector" by distributing shear loads from the floor or roof assembly above and transferring them along the length of the wall to sheathed corners or braced wall segments. When the rafters or joists are directly in line with the studs of the wall below, only a single top plate is needed. Double top plates, however, add strength and redundancy to the structure and may be required in high-wind and seismic areas.

HEADER HANGERS support the weight of the header and transfer its gravitational load into the king studs. They replace jack studs.

Load path highlighted in red

OSB or plywood **SHEATHING** on the exterior of the wall resists uplift on the top plate and, as a shear-resisting element, keeps the wall from racking. In some cases, the midwall sheathing is replaced with nonstructural rigid-foam insulation. The corners of the house are then reinforced with plywood or fabricated shear walls.

KING STUDS prevent the header from rotating within the assembly and transfer wind and dead loads to the roof framing above or floor framing below.

The **MUDSILL** is a rot-proof member that attaches the frame to the foundation.

The **RIM JOIST** transfers loads from the bottom of one wall to the top of a lower wall, or from the bottom of a first-floor wall to the foundation. The rim joist also keeps the ends of the floor joists from racking and transfers shear loads from the floor framing to the top plates or to the foundation wall. If properly designed, the rim joist also can act as the window and door header for the wall below. This is usually done only with smaller openings. Care should be exercised to ensure that splices in the rim joist do not end up over these openings.

Framing with a Crane

BY JIM ANDERSON

The other day, I kept an eye on the three-man framing crew working across the street. Two of the guys spent the day in the mud, hauling material for the second floor into the house. When my crew was rolling up for the day, the other crew had carried most of the lumber inside the house but still needed to pull it up to the second floor. It cost this crew about $300* in labor just to get the studs in the door.

I'd have hired a crane for this job and moved all that lumber in about an hour, which would have cost me $125. Even if there were no money savings, saving wear and tear on my body and those of my crew would make the crane worthwhile.

I started to think of better ways to use a crane six years ago when my brother and I went into business framing houses. We carried most of the material the hard way, but we hired a crane to set the steel I-beams.

It occurred to us that although the crane company charged us for a full hour of crane time, setting those three steel beams took 20 minutes. With that realization, we decided to fill the other 40 minutes of that hour by using the crane as our laborer.

My brother has moved on, and I now have my own two-helper crew. I still call in a crane several times

for an average house. Where I work, in the suburbs south of Denver, Colorado, cranes are pretty common. The ones that I hire usually have no move-in fee, just a one-hour minimum charge of about $125*.

Preparation Is Key

Before the crane arrives, I try to ensure that the lumber is dumped fairly close to the house, but not where it will be in the crane's way. When the crane does arrive, I discuss the sequence of the lift with the operator and crew. For efficiency, everybody has to know what's coming next.

One crew member stays near the material to rig it to the crane. The other two stay near where the material will be installed. To avoid confusion, one of these carpenters is the designated signaler (see the sidebar on p. 60), whereas the other jockeys the load into position.

Most of the houses that my crew and I frame have three to five steel beams holding up the first floor. These beams are the first things that the crane sets. We either have the Lally columns cut to length or have ready temporary posts. Long 2×4s are on hand to brace the I-beam to the mudsills once it's in place.

We set the beam that's farthest back in the building first, then move sequentially to the front. This

A WELL-RIGGED JOB GOES SMOOTHLY AND SAFELY

THREE KINDS OF RIGGING EQUIPMENT GET US through any house. The most frequently used are nylon straps, followed by steel cables and four-chain rigging.

NYLON STRAPS ARE RIGGED IN TWO WAYS

CRADLE-RIGGING RUNS THE STRAP UNDER THE LOAD IN A U-SHAPE. Great for trusses and walls, they should never be used to lift studs or joists overhead.

CHOKE-RIGGING TIGHTENS AS THE CRANE LIFTS. Choke rigs limit movement within lifts such as studs or sheathing, and they keep beams from slipping out of single straps. To avoid excessive flexing, I-joists (right) are choked in about one-quarter of the joists' length from each end.

CABLES AND FOUR-CHAIN RIGS ARE LESS COMMON

FOUR-CHAIN RIGS RAISE THE ROOF. Hooked to the crane with a ring, each chain can be snubbed to a length that will enable a preassembled truss rack to fly level.

STEEL CABLES CAN DAMAGE THE EDGES OF MATERIAL. They're limited to raising single trusses and sliding under material dropped directly onto the ground so that it can be raised enough to get a strap below.

process avoids swinging anything over set beams, eliminating the chance of dropping one beam and taking out two.

We stack any scrap from the basement on a piece of sheathing. We lift this scrap out of the hole with the crane and swing it right over to the Dumpster®.

Any built-up wood or laminated veneer lumber (LVL) beams are nailed together, and any necessary hangers are installed beforehand. As we work back to front, I set these beams in sequence with the steel.

The stacks of floor joists are next, and after they have all been set on the foundation walls, we move as much of the material as we can closer to the house. We put the lifts of floor sheathing on top of

2×4 stickers about 3 ft. from the front of the house, allowing room to work but putting it within easy reach. Then we use the crane to move the rim-joist material to the top of the stacks of sheathing. We cut the rims here, using the sheathing stacks as 1,500-lb. sawhorses.

Have the Garage Walls Ready to Lift

Most of the houses we frame have three-car garages with one single door and one double door. A typical garage-door wall is 30 ft. long and 10 ft. tall, with an 18-ft. and a 9-ft. double-LVL header. I never want to lift a wall like this by hand. Before the crane arrives, we frame and stand the sidewalls that we can easily lift by hand. We also frame the front wall but leave it on the ground for the crane to lift.

The crane lifts the wall and swings it to the garage foundation. As it gets close, we guide the anchor bolts into the holes we've drilled for them in the bottom plate, then down the wall. Now the crane holds the wall in place as we nail the corners, tie in the plates, and nail on some braces. With the wall set and braced, we give the "all finished" signal and send the crane home. This entire lift—I-beams to garage wall—usually takes only about an hour.

THE STEEL IS READY TO GO WHEN THE CRANE ARRIVES. The wood sills to which the joists will be nailed are already attached to the steel, and the joist locations are marked.

GETTING THE CRANE TO DO WHAT YOU NEED

THERE ARE STANDARD HAND SIGNALS that all crane operators and the people who hire them should know. In addition, three rules and a suggestion can make communication a sure thing.

1. Keep your signal in one place.
2. If you can't see the operator through a maze of studs, trusses, or bracing, the operator can't see you. Make eye contact, and then make your signals in front of your face.
3. If your gloves and clothing are of similar colors, make your signals away from your body where the operator can see. Another way to communicate with the operator is with two-way radios.

TO RAISE OR LOWER A LOAD, POINT UP OR DOWN AND ROTATE YOUR FINGER. To move the load slowly, put your opposite hand above or below your signal, as if you're pointing at your palm. When the load is down, a quick circle with the hand signals that all is clear and that the hook can be dropped to free the load.

RAISING OR LOWERING THE BOOM MOVES A LOAD TOWARD OR AWAY FROM THE CRANE. Thumb up moves the load toward the crane. Thumb down moves the load away. Either gesture pointed at the opposite palm means to go slowly. To move the load without raising or lowering it, point your thumb up or down while opening and closing your fist.

SWINGING THE BOOM IS AS SIMPLE AS POINTING WHERE YOU NEED TO MOVE THE LOAD. Finally, the most important signal is a closed fist, for stopping the crane. Additional signals are used to guide the larger cranes used on commercial jobs, but these six signals and their variations should get you through most residential work.

A CRANE MAKES SHORT WORK OF RAISING TOP-HEAVY GARAGE WALLS. Once the wall is set over the anchor bolts, the crane steadies the wall until it's tied to the others and braced.

Better Than Carrying Material to the Second Floor

After we've framed the first-floor walls, we build a section of the second floor to serve as a staging area for our next lift. This area is usually a corner that's, say, 500 sq. ft. to 800 sq. ft. We also frame most of the walls around this area before placing any material here. This prep work saves having to move lumber to make room to frame the walls.

Once this staging area is done, I call the crane again to lift the studs, the plates, and the sheathing for the second-floor walls, as well as the balance of the second-floor joists and beams.

Full lifts of studs or of OSB sheathing are pretty heavy loads that should never be set in the middle of a joist span. We set lifts of material on 2×4s to spread the weight and to leave room to remove the straps once the crane lets go of the load. We always set our material on or near the main bearing beams and walls below the floor. If for some reason a load must be set midspan, we split it into smaller bundles that can be

BE PREPARED FOR THE CRANE'S SECOND VISIT. This second-floor staging area is framed to provide storage space for crane-lifted studs and sheathing. Because stored material would otherwise be in the way, the author also frames the walls before the crane arrives.

spread out. We sometimes build a temporary wall below the floor to help spread the load to more joists.

During this lift, we set any second-floor steel or built-up LVLs. Commonly, there is a beam that runs between two walls, each end supported by posts made of studs or by Lally columns enclosed within pockets in the wall. Although we frame these walls with their top plates uninterrupted, we leave the posts out until we set the beam.

To set a beam that runs between walls, we pull down one end and slide it far enough into its pocket that the beam's other end clears the wall plate. Then we have the crane operator lower the beam, and we seesaw it into place. With the crane snugging the beam to the underside of the top plates, we tip in the supporting columns.

Raising Balloon-Frame Walls

The only first-floor walls we don't lift by hand are any tall, balloon-frame walls, such as those of rooms with very high ceilings. We frame and sheathe these walls with the other walls but leave them flat until the crane comes again. For safety alone, lifting these walls is worth hiring a crane.

With the wall's bottom plate on its layout line, we secure it to the deck about every 4 ft. using pieces of the steel strap that bands lifts of lumber. This strap acts as a hinge, keeping the wall from kicking out during the lift.

We stop the lift about a third of the way up to attach the braces that will steady the raised wall. These braces go about two-thirds of the way up the wall. Once the braces are nailed to the wall, we stand it up

WITH THE WALLS PARTIALLY RAISED AND SHORED FOR SAFETY, the braces are tacked so that they can pivot as the wall rises.

TWO STRAPS ARE USED TO RAISE TALL WALLS. Had there been no windows, the crew would have cut small holes in the sheathing for the straps.

the rest of the way, then plumb and brace it before unhooking the crane.

The Final Lift Sets the Roof Trusses

With the second-floor walls plumbed and lined, it's time for the crane again. This time it will lift our sheathing, roof-framing lumber, prebuilt truss racks, and any single trusses. First, we swing up the roof sheathing and set it in three or four spots on the second floor. Then we send up the lumber for framing dormers and valleys. This lumber is usually set in the main hallway of the upper level, where we have room to maneuver long pieces up into the roof framing.

When the site and truss design allow, we assemble, sheathe, and brace the trusses into 6-ft. to 18-ft. roof sections on the ground. We can set a simple gable roof, preassembled into two to three sections, in about a half-hour.

Our total crane time on an average house is about 4½ hours. In most cases, we hire a 3-ton crane. These cranes have 90-ft. booms that will reach about 60 ft. with most loads that we see in residential construction. Only once have we needed a larger crane. If you aren't sure of the size crane that you need, call the crane company and describe your lift. They'll know.

There are two other advantages to hiring a crane. The first is safety. We once helped another framer attempt to raise a large wall by hand. It got away from us, pinning one of my employees below. Added to his pain and disability is the fact that my worker's comp premiums went up 50 percent.

The second advantage is job-site security. Using a crane to lift material into the house as soon as possible makes it less accessible to thieves. They'll likely head across the street to the house where the lumber is still sitting where it was delivered, conveniently right next to the curb.

*Prices are from 2001.

CRANE SAFETY IS MOSTLY COMMON SENSE

Cranes offer immediate safety benefits by lifting heavy loads that you might otherwise attempt to manhandle. And by reducing the repetitive toting and lifting of construction, they can foster long-term benefits in the form of healthy backs, knees, and shoulders.

However, cranes bring with them some danger, simply because they carry heavy things overhead. Obeying three rules should get you home safe after every day of crane work.

- Wear a hard hat: OSHA requires it.
- Don't stand under suspended loads.
- Don't become trapped between the load and a wall or a drop-off. A sudden horizontal movement of the load could crush you or send you flying.

All About Headers

BY CLAYTON DEKORNE

Like many carpenters in the Northeast, I was taught to frame window and door headers by creating a plywood-and-lumber sandwich, held together with generous globs of construction adhesive and the tight rows of nails that only a nail gun could deliver. Years later, I learned my energetic efforts to build a better header were an exceptional waste of time and resources. Neither the plywood nor the adhesive contributed much strength, only thickness, and this perfect thickness helped only to conduct heat out of the walls during the severe winters common to the region.

At the same time I was laying up lumber sandwiches, young production framers on the West Coast were framing headers efficiently using single-piece 4×12s. They needed only to be chopped to length and filled the wall space above openings, eliminating the need for maddeningly short cripple studs between the top of the header and the wall plate. Nowadays, however, such massive materials are relatively scarce and remarkably expensive, even on the West Coast. So although solid-stock headers certainly save labor, they no longer provide an economical alternative.

With these experiences in mind, I set out to discover some practical alternatives, surveying a number of expert framers in different regions of the

THE HEADER YOU CHOOSE. This decision affects not just strength, but also cost, energy efficiency, and drywall cracking.

country. Header framing varies widely from builder to builder and from region to region. Even when factors such as wall thickness and load conditions are made equal, building traditions and individual preferences make for a wide range of header configurations. The examples shown here are just a few of the options possible when you mix and match features, notch cripple studs, and sift in engineered materials. But they aptly demonstrate a number of practical considerations that must be kept in mind when framing a good header.

Big Headers Need More Studs

A header transfers loads from the roof and floors above to the foundation below by way of jack studs (see the sidebar on p. 66). This means the header not only must be deep enough (depth refers to the height of a beam: 2×10s are deeper than 2×6s) for a given span to resist bending under load, but also must be supported by jack studs on each end that are part of a load path that continues to the foundation.

The *International Residential Code (IRC)* specifies not only header size but also the number of jack studs for most common situations. Although most windows and doors require just one jack stud at each end, long spans or extreme loads may call for two or more jack studs to increase the area bearing the load. If the loads on any header are concentrated over too small an area, the wood fibers at the ends of the header can be crushed. This can cause the header to drop, which in turn can crack drywall or, particu-

larly with patio doors and casement windows, cause the door or window to jam.

Header hangers, such as the Simpson Strong Tie® HH Series (see the bottom sidebar on p. 67), can be used to eliminate jack studs altogether. I've used them in some remodeling situations when I needed to squeeze a patio door or a wide window into an existing wall that didn't have quite enough space for double jack studs. One jack and a Simpson HH Series hanger did the trick.

How Big a Header Do You Need?

Unless you're an engineer, the easiest way to size headers built with dimensional lumber is to check span charts, such as those in the IRC. The old rule of thumb is that headers made of double 2× stock can span safely in feet half their depth in inches. So by this rule, a double 2×12 can span 6 ft.

However, header spans vary not only with size, but also with lumber grade and species, with the width of the house, with your area's snow load, and with the number of floors to be supported. Consequently, the IRC provides 24 scenarios in which that double 2×12 header can span a range from 5 ft. 2 in. to 9 ft. 9 in. Check the code.

The Trouble with Cripples

Header size often is based on factors other than strength requirements. Many framers purposely oversize headers to avoid filling the space between the header and the double top plate with short studs (cripple studs, or cripples). In a nominal 8-ft.-tall wall, a typical cripple stud measures 6 in. to 7 in. Such short studs are ungainly and are prone to splitting when they are nailed in place. Yet a double 2×12 header can be tucked beneath the double top plate, filling this miserable space and creating a proper opening for common 6-ft. 8-in. doors. Alternatively, builder John Carroll relies on a double 2×10 header with a 2×6 nailed flat along the bottom edge, which provides nailing for the head trim in a 2×6 wall.

However, such deep headers are oversize and add considerable cost, not to mention waste wood. Most

LIKE A BRIDGE OVER A RAVINE, A HEADER SPANS A WINDOW OR DOOR

HEADERS ARE SHORT BEAMS THAT typically carry roof and floor loads to the sides of openings for doors or windows. Jack studs take over from there, carrying the load to the framing below and eventually to the foundation. That's called the load path, and it must be continuous. The International Residential Code (the most common code nationwide) has a lot to say about headers, including the tables you need to determine the size header required for most situations. If it's not in the IRC, you need an engineer.

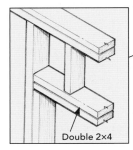

Double 2×4

Where You Don't Need a Structural Header

In a gable-end wall that doesn't support a load-bearing ridge, or in an interior nonbearing partition, headers up to 8 ft. in length can be built with the same material as the wall studs. Inside, a single 2× is often sufficient, although a double 2× is helpful for securing wide trim. In a gable-end wall, however, a double 2× is needed to help resist the bending loads exerted by the wind.

← Load path must be continuous from roof to foundation.

Double rim joist as header

Hidden Header

Here's a wood-saving trick. By simply adding an additional member to it, the rim joist above an opening can serve as a header. Two caveats: You'll need to use joist hangers to transfer the load to the header. And if a standard header of this size requires double jack studs, so does a double rim joist.

Cripple Studs Fill the Void

If the header doesn't fill the space all the way to the top plate, cripple studs are used to carry the load from the rafters, joists, or trusses above to the header below. For nonbearing headers, the IRC requires no cripples if the distance to the top plate is less than 24 in.

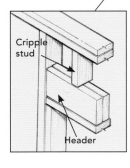

Cripple stud

Header

King stud

Header

Jack stud

Kings High, Jacks Low

King studs are the same height as the wall studs, running plate to plate. Nails driven through them into the header's end grain stabilize the header. Jack studs are shorter and fit below the header to carry loads downward. Because longer-spanning headers usually carry greater loads, you may need an extra jack under both ends of big headers. Check your building code.

INSTALLATION GUIDELINES

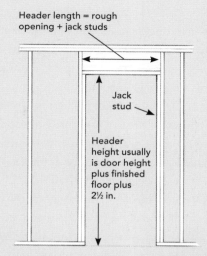

Header length = rough opening + jack studs

Jack stud

Header height usually is door height plus finished floor plus 2½ in.

TYPICALLY, HEADER HEIGHT IS ESTABLISHED BY the door height, and window headers are set at this same height. In homes having 8-ft. ceilings, a header composed of 2×12s or of 2×10s with a flat 2×4 or 2×6 nailer on the bottom accommodates standard 6-ft. 8-in. doors, as shown in the illustration at right.

In a custom home with cathedral ceilings and tall walls, however, header heights can vary widely. And if the doors are a nonstandard height, you'll need to figure out the header height. Finding the height of the bottom of the headers above the subfloor is a matter of adding up the door height, the thickness of the finished-floor materials, and 2½ in. (to allow space for the head jamb and airspace below the door). There are exceptions. Pocket doors typically require a rough opening at least 2 in. higher than a standard door. Windows may include arches or transoms, which affect the rough opening's height.

To find the header length for windows, add 3 in. to the manufacturer's rough-opening dimen-

sion if there is to be one jack stud on each side, or 6 in. if two jacks are called for. For doors with single jack studs, add 5½ in. to the door width to allow for jack studs, door jambs, and shim space. If double jacks are needed, then the header should be 8½ in. longer than the door width.

These guidelines follow one fundamental rule of framing rough openings: Know your windows and doors. If you don't have the window or door on site, at the very least check the manufacturer's catalog to verify the rough-opening dimensions. Don't rely on the plans alone, and when in doubt, call the manufacturer.

HANG YOUR HEADER

SOMETIMES, PARTICULARLY IN REMODELING, there just isn't room for a jack stud. The IRC permits header hangers, such as Simpson's HH4 for 2×4 walls and HH6 for 2×6 walls, to substitute for single jack studs. These hangers are spiked with 16d common nails to the king stud.

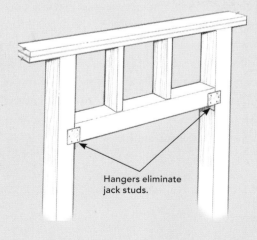

Hangers eliminate jack studs.

window and door openings are only 3 ft. or so and might only require 2×6 headers. But perhaps the biggest drawback of wide lumber is that there's more of it to shrink. Framing lumber may have a moisture content of 19 percent. Once the heat is turned on, lumber typically dries to a moisture content of 9 percent to 11 percent, shrinking nominal 2×10s and 2×12s as much as ¼ in. across the grain. On the other hand, 2×6s might shrink only half that.

Shrinkage reduces the depth (or height) of the header; because the header is nailed firmly to the double top plate, a gap usually opens above the jack studs. As the header shrinks, it tends to pull up the head trim, which has been nailed to it, opening unsightly gaps in the casing and cracking any drywall seam spanning the header. The gap above the jack stud now means the header isn't supporting any load—until the first wet snowfall or heavy winds bring a crushing load to bear on the wall and push the gap closed, causing the top plates to sag, which can crack the drywall in the story above.

Shrinkage can be reduced using drier lumber, preferably at about 12 percent. However, lumber this dry may be difficult to find unless you can condition it yourself. As an alternative for spanning a large opening, consider using engineered materials (see the sidebar on pp. 70–71). Laminated-veneer lumber or parallel-strand lumber (PSL) shrinks much less than ordinary lumber.

If wide dimensional lumber is unavoidable, structural engineer Steve Smulski suggests that cracking can be minimized by not fastening the drywall to the header. This way, the header moves independently of the drywall, which then is less likely to crack. To prevent trim from moving as the header shrinks, attach the top piece of trim to the drywall only, using a minimal number of short, light-gauge finish nails and a bead of adhesive caulk.

SAWN-LUMBER HEADERS

WHEN IT COMES TO SAWN-LUMBER headers, traditional materials still carry the load.

DOUBLE 2×6 HEADER

Fine Homebuilding contributing editor Mike Guertin, whose day job is building houses in Rhode Island, uses the smallest allowable header depth to span the opening. Although he must toenail cripples above each header, he argues that this header is the most economical. For starters, it conserves lumber. It also reduces the area of solid material in the wall, thus reducing thermal bridging. Although the area is kept to a minimum, Guertin is also careful to keep the header to the outside of the wall, providing a gap that may be insulated with foam or wet-spray cellulose when the rest of the wall is insulated. A 2×3 nailed to the lower edge of the header provides attachment for trim.

DOUBLE 2×10 HEADER

A common header variant is used by North Carolina builder John Carroll. Built from double 2×10s, a stud-width nailer flat-framed along the bottom edge eases attaching sheathing or trim. Because this header is less than the full thickness of the wall, it allows for a piece of ½-in. foam to add a bit of insulation.

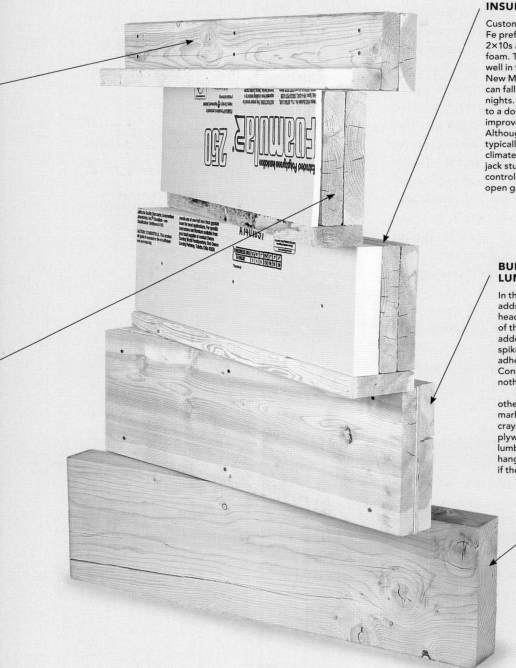

INSULATED HEADER

Custom builder David Crosby of Santa Fe prefabs insulated headers from 2×10s and 2-in. extruded polystyrene foam. This option works particularly well in the cold mountains of northern New Mexico, where air temperatures can fall well below zero on winter nights. Even adding some ½-in. foam to a double header in a 2×4 wall improves the thermal performance. Although lumber in New Mexico is typically quite dry due to the arid climate, Crosby ties the header to the jack stud with metal framing plates to control header shrinkage that could open gaps in the trim.

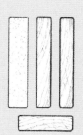

BUILT-UP PLYWOOD AND LUMBER

In this header sandwich, plywood adds only thickness so that the header will fit flush to each face of the wall. There is little strength added, even if the header is spiked together with construction adhesive between each layer. Construction adhesive adds nothing to the strength of a beam.

Before assembling this (or any other header), crown the lumber, marking it clearly with a lumber crayon, and keep the crown up. Rip plywood ½ in. narrower than the lumber to prevent the pieces from hanging over the edges, especially if the lumber has a crown.

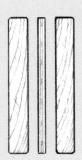

SOLID-STOCK HEADER

Once standard fare for West Coast production framers, a solid header made with a single 4×12 tucks tight under the top plates in a wall, eliminating the need for short cripple studs. Although this option saves substantial labor, the availability of full-dimension lumber is limited mainly to the West Coast. Even there, solid-stock headers are expensive and may not be cost-effective unless the opening requires the load-bearing capacity of such large-dimension stock.

Avoiding Condensation

In cold climates, uninsulated headers can create a thermal bridge. According to Smulski, the uninsulated header makes the wall section above windows and doors significantly colder than the rest of the wall. When the difference between the inside and outside air temperatures is extreme, condensation may collect on these cold surfaces, and in the worst cases, mold and mildew may begin to grow.

To avoid condensation, it's important that any uninsulated header doesn't contact both the sheathing and the drywall. Unless you're building 2×4 exterior walls using full-thickness headers such as solid lumber or ones built out to 3½ in. with plywood, avoiding this situation is simple. Keep the header flush to the outside of the framing so that it contacts the sheathing. Because most other types of headers are narrower than the studs, there will be some airspace between the header and the drywall, which makes a dandy thermal break. In cold climates, a 2×10 insulated header, like the one used by David Crosby of Santa Fe, New Mexico, works well (see the sidebar on pp. 68–69). Another option that avoids solid lumber is a manufactured insulated I-beam header.

SOURCES

APA ENGINEERED WOOD ASSOCIATION
Nailed Structural-Use Panel and Lumber Beams
www.apawood.org

SIMPSON
www.strongtie.com

TRUSJOIST'S® PARALLAM®
www.woodbywy.com

ENGINEERED-WOOD HEADERS

ENGINEERED-WOOD HEADERS cost more, but they do more, too.

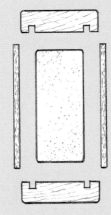

STORE-BOUGHT INSULATED HEADERS

Essentially a double-webbed I-joist with a chunk of rigid foam wedged in the middle, these engineered headers offer insulation, strength, and light weight.

PARALLEL-STRAND LUMBER

Parallel-strand lumber, such as TrusJoist's Parallam, is available as stud-width stock. Performing much like LVL, parallel-strand header stock is pricier than solid sawn lumber but 1½ times as stiff and 3 times as strong.

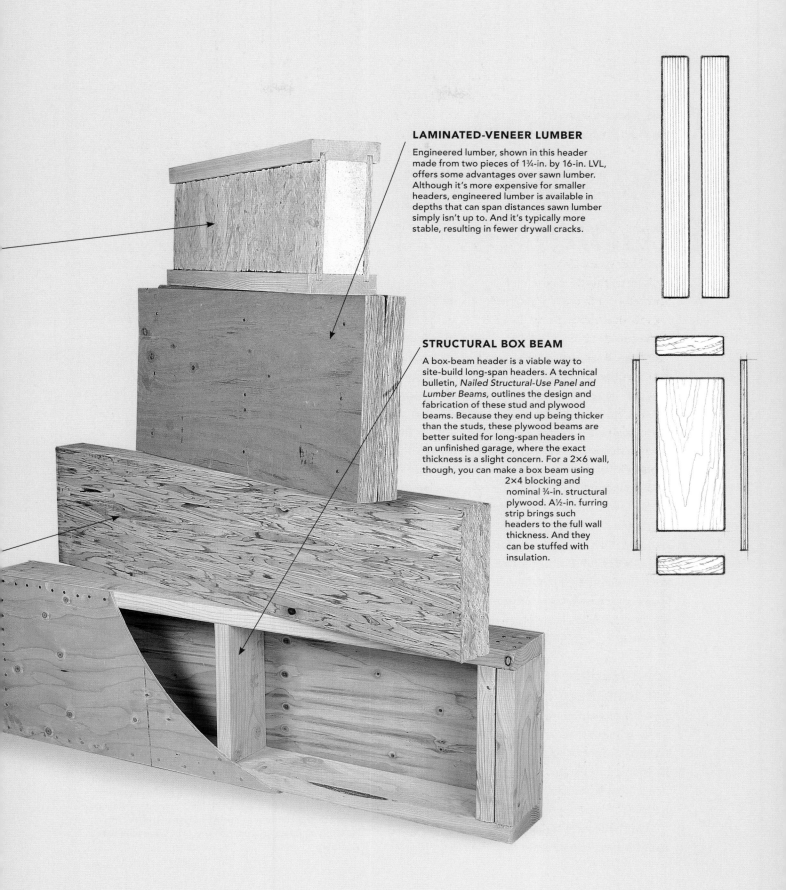

LAMINATED-VENEER LUMBER

Engineered lumber, shown in this header made from two pieces of 1¾-in. by 16-in. LVL, offers some advantages over sawn lumber. Although it's more expensive for smaller headers, engineered lumber is available in depths that can span distances sawn lumber simply isn't up to. And it's typically more stable, resulting in fewer drywall cracks.

STRUCTURAL BOX BEAM

A box-beam header is a viable way to site-build long-span headers. A technical bulletin, *Nailed Structural-Use Panel and Lumber Beams*, outlines the design and fabrication of these stud and plywood beams. Because they end up being thicker than the studs, these plywood beams are better suited for long-span headers in an unfinished garage, where the exact thickness is a slight concern. For a 2×6 wall, though, you can make a box beam using 2×4 blocking and nominal ¾-in. structural plywood. A ½-in. furring strip brings such headers to the full wall thickness. And they can be stuffed with insulation.

Common Engineering Problems

BY DAVID UTTERBACK

Over the past 22 years as a builder, building inspector, and lumber-industry representative, I've inspected a great deal of framing in all parts of the country. Terms and techniques vary from region to region, but mistakes don't. The same problems tend to show up over and over.

Here, I'll examine some of these problems from an engineering standpoint and look at what can be done to avoid them. All these situations are addressed in similar ways by each of the three major building codes. Before going further, I must emphasize that difficult framing problems often require complex engineered solutions. When the going gets tough, your best bet is to enlist the services of a good engineer. It's a lot cheaper than defending yourself in a lawsuit.

Joist-Hanger Nails Are Not Meant for Installing Joist Hangers

Joist hangers are marvelous devices for supporting joists or beams that cannot rest directly atop vertical framing members. To get the most structural capacity from a joist hanger, you must use the correct hanger for the joist and place the right nail in every nail hole.

Many builders mistakenly assume that the right nails for every situation are the 1½-in. long "joist-

BRACING FOR BAD NEWS. Placing too big a pile of sheathing on an unbraced truss roof can lead to disaster. Amazingly in this case, the falling dominos were halted when a quick-thinking carpenter was able to brace the remaining upright trusses before the collapse reached the part of the roof where he was working.

hanger nails" sold by the manufacturer. In truth, these nails are intended for anchoring the sides of the hanger to a single joist without piercing the other side.

Joist-hanger nails have the same diameter, and therefore the same shear capacity, as 10d common nails, but their shorter length gives them less withdrawal resistance. For maximum strength, nothing smaller than 10d common nails (or 16d sinkers, which have the same diameter) should be used to attach a single joist hanger to a beam. To attach a double joist hanger to a beam, 16d commons should be used.

This fact does not mean that you can never use the short nails to support a joist hanger. But if you do, you must reduce the load. If joist-hanger nails are used instead of 10d commons to support a single hanger, you can use only 77 percent of the load value of that hanger. If they are used instead of 16d commons to support a double hanger, the load capacity drops to 64 percent. It's always wise to check with the hanger manufacturer if you are not sure what size nails to use. Some hangers have the required nail size stamped directly on the hanger.

Besides nails, you also need to understand the differences between hangers. Some hangers have little dog ears on the side of the hanger sticking out at 45-degree angles (I've seen framers bend the ears over to get them out of the way). These hangers require what is called double-shear nailing: Common nails are driven through these holes at an angle into the joist and on into the supporting beam or header, distributing the load through two points on each joist nail for greater strength. If you use this type of hanger, make sure it is nailed correctly.

Load-Bearing Cantilevers Need Careful Engineering

Many builders lay out cantilevers according to a simple rule of thumb: "One out, two in." This rule means that for whatever length the joists extend past their bearing point, they should run back in at least twice as far. Although technically correct, the rule applies to non-load-bearing applications only, and even then has its limits. In nonbearing applications, a joist may not cantilever more than four times its depth. Therefore, a 2×10 joist should cantilever no more than 37 in. (4 × 9¼ in.), regardless of its length.

Non-load-bearing cantilevers can include sun decks and even bay windows (the cantilever supports only the weight of the window; any loads above are carried on a header set into the main wall). On the other hand, a zero-clearance fireplace with a two-story wood-frame chase would impose a significant bearing load on a cantilever.

JOIST-HANGER NAILS ARE TOO SHORT FOR SOME APPLICATIONS

The 1½-in.-long joist-hanger nails are primarily intended to attach the hanger to the sides of a single joist. For maximum strength, full-length 10d common nails (or 16d sinkers) should be used to fasten the joist hanger to the beam.

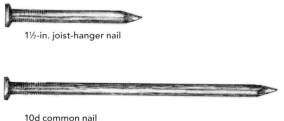

1½-in. joist-hanger nail

10d common nail

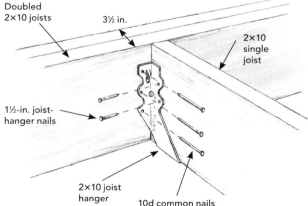

Doubled 2×10 joists

3½ in.

2×10 single joist

1½-in. joist-hanger nails

2×10 joist hanger

10d common nails

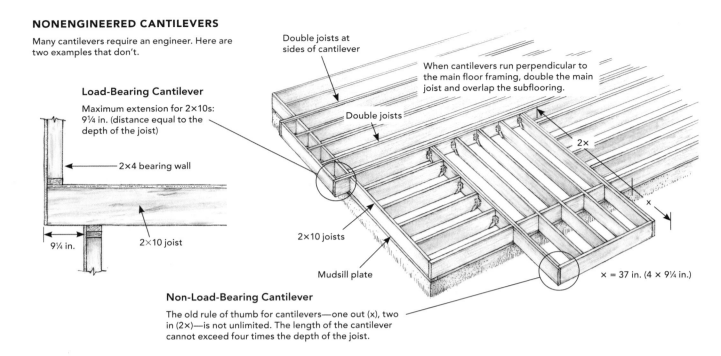

NONENGINEERED CANTILEVERS

Many cantilevers require an engineer. Here are two examples that don't.

Double joists at sides of cantilever

When cantilevers run perpendicular to the main floor framing, double the main joist and overlap the subflooring.

Double joists

2×

Load-Bearing Cantilever

Maximum extension for 2×10s: 9¼ in. (distance equal to the depth of the joist)

2×4 bearing wall

2×10 joist

9¼ in.

2×10 joists

Mudsill plate

x = 37 in. (4 × 9¼ in.)

x

Non-Load-Bearing Cantilever

The old rule of thumb for cantilevers—one out (x), two in (2×)—is not unlimited. The length of the cantilever cannot exceed four times the depth of the joist.

Some simple load-bearing cantilevers can be built without paying an engineer. Because loads transfer through solid-sawn joists at 45-degree angles, codes allow load-bearing cantilevers that extend the same distance as the joists are wide. In other words, you could set a bearing wall on the end of 2×10 floor joists that are cantilevered 9¼ in. without risking a correction notice.

Cantilevered joists that run perpendicular to the main floor joists may have another problem: If the connection between the two is not constructed properly, a teeter-totter effect could force the inside edges of the cantilevered joists upward, creating a hump in the floor.

To prevent this unpleasantness, the cantilevered joists should butt into a main joist that has been doubled to serve as a header. The connection between the cantilevered joists and the header should be securely constructed to prevent independent movement. As an added measure, subflooring should overlap the joint where the cantilevers and the main joist meet.

Bearing Walls Should Line Up with Their Supports

To transmit loads smoothly from roof to foundation, bearing walls must be stacked closely above one another. Where I-joists are involved, each bearing wall must sit directly over the top of its support because the web of an I-joist has little cross-sectional strength. Solid blocking or squash blocks also need to be installed according to the manufacturer's instructions to carry the load around the web and prevent the web from buckling.

Solid-sawn floor joists have more cross-sectional strength than I-joists, which allows you a little bit of leeway if you need to offset a bearing wall from its support. You can basically treat this situation the same as you would a load-bearing cantilever, meaning you could offset the bearing walls the same distance as the depth of the floor joist. If you had a 4½-in.-wide flange supporting 2×10 floor joists, you could set a bearing wall 9¼ in. to each side of the beam's edge and still meet code, giving you almost 2 ft. to play with.

To prevent rotation of the joists, the codes also require full-depth solid blocking over beams or over

SUPPORTING BEARING WALLS

When solid-sawn floor joists are used, bearing walls may be offset a distance equal to the depth of the joist. To keep the joists from rolling over, full-depth solid blocking is required between the joists where they rest on the bearing walls.

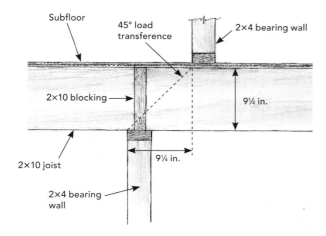

Subfloor

45° load transference

2×4 bearing wall

2×10 blocking

9¼ in.

2×10 joist

9¼ in.

2×4 bearing wall

AN OTHERWISE WELL-BUILT HOUSE. Because the tiny return walls on each end of the three-car garage could offer little resistance, lateral loads from an earthquake literally twisted this house off its foundation. The structural integrity of the rest of the house was largely unaffected by the earthquake.

WIDER IS BETTER

Conventionally framed garage return walls must be wide enough to resist lateral loads imposed by high winds or ground movement. The Uniform Building Code minimum width of 2 ft. 8 in. can be reduced, however, if precise shear-wall schedules are followed. The following is one example of a 16-in. shear wall.

bearing walls that support floor joists. As lateral loads, such as wind, are placed on the building, they're transferred into the floor diaphragm through the joists on their way to the foundation. By themselves, the nails that attach the plywood subfloor to the joists do not have the strength to resist these forces; if the floor joists are not blocked, they could actually roll over and end up lying flat.

Garage Walls and Cripple Walls Need Extra Bracing

Most regions of the country aren't threatened by earthquakes, but nearly every place is exposed to high winds. It is extremely important that walls be properly braced to resist these lateral loads, or the results could be catastrophic. Builders can generally rely on structural sheathing to brace walls, but that's not always enough.

Among the weakest points in a house frame are the narrow return walls on the sides of the garage door. Tall narrow walls are inherently difficult to brace properly against high lateral loads; this fact is why the Uniform Building Code (UBC) now re-

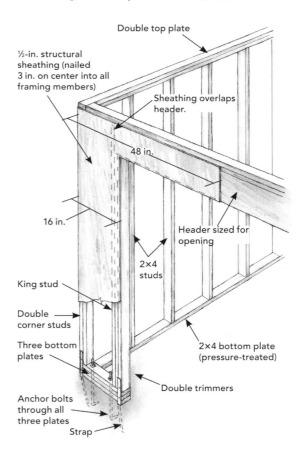

Double top plate

½-in. structural sheathing (nailed 3 in. on center into all framing members)

Sheathing overlaps header.

48 in.

16 in.

Header sized for opening

2×4 studs

King stud

Double corner studs

Three bottom plates

Anchor bolts through all three plates

Strap

2×4 bottom plate (pressure-treated)

Double trimmers

quires a minimum 2-ft. 8-in. width for garage return walls. If you absolutely must squeeze in space for three cars, you can build a shear wall on site—by following a precise schedule for framing, nailing, and bolting—that will allow you to reduce this width to 24 in. or possibly even 16 in.

Cripple walls (short kneewalls that run between the mudsills and the first-floor joists) are another weak link in the structural chain. Besides transferring vertical loads through to the foundation, these walls must also resist lateral loads. Cripple walls are effectively shear walls, and as such, they must be braced with structural panels and nailed 6 in. on center to provide the shear resistance necessary to support the structure above.

Another bracing point often overlooked is the connection between first-story and second-story walls. Most of the time, builders brace these walls independently of each other. Because lateral loads such as high wind can impose torque (turning or twisting energy) on a building, the upper story will move more than the lower story if the two aren't tied together.

Fortunately, these walls can be tied together easily. One solution is to overlap the sheathing panels between floors (blocking the panel edges may be neces-

sary in areas that are subject to high lateral loads). If sheathing is already in place, another solution is to tie the lower studs to the upper ones using metal straps specifically manufactured for that purpose.

Think Twice before Cutting Beams

It's easy to pull out a saw and cut off the top corner of a beam that must be kept beneath a roofline. But if too much cross section is removed, shear forces can cause the beam to split and eventually to fail. For solid-sawn beams, you should leave at least half the width of the beam above the supporting wall and confine the length of the tapered cut to no more than three times the original width of the beam. If you don't have room to leave this much cross section, your best bet is to lower the beam (set it in a pocket) or have a tapered beam engineered.

Solid-sawn beams may be notched one-quarter their depth at the ends and one-sixth their depth in the outer thirds of the span. Holes may be drilled in a beam from face to face, but never from edge to edge. The diameter of the hole may be as big as one-third the depth of the beam, but it must be at least 2 in. from the top or bottom edge.

WHERE TO CUT OR DRILL BEAMS

Plumbers, electricians, and HVAC installers are as guilty as carpenters when it comes to carving—and weakening—beams. Here's a brief rundown on what the codes allow.

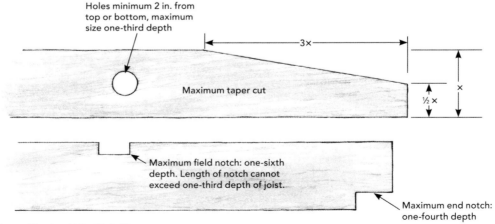

Holes minimum 2 in. from top or bottom, maximum size one-third depth

3x

Maximum taper cut

½ x

x

Maximum field notch: one-sixth depth. Length of notch cannot exceed one-third depth of joist.

Maximum end notch: one-fourth depth

No notching in middle third of beam

Rafter Ties Must Be Near the Plates to Be Effective

Many builders confuse collar ties with rafter ties. Both are horizontal framing members that connect rafters, but that's where the similarities end. Collar ties (which are required by the Southern Building Code and no other) function to resist the pressures of wind uplift on a roof by holding the rafters together where they meet the ridge. As high up as they are, collar ties have no leverage to prevent the rafters and walls from spreading outward. That job is best done by the ceiling joists.

If there are no ceiling joists or if the joists run perpendicular to the rafters, then the code requires rafter ties. Similar to a ceiling joist, a rafter tie is typically a 2×4 that runs parallel to the rafters, from outside wall to outside wall, and ties the rafters together as close to the top plate as possible. Rafter ties need to be installed every 4 ft. down the length of the roof.

Rafter ties do not have to be at ceiling height to be effective, but they must not be placed any higher than the lower third of the roof pitch. In other words, measure vertically from the outside wall's top plate to the bottom of the ridge and place the rafter ties within the lower third of that measurement. Once they get above that point, they lose their most effective leverage.

I've seen builders compound their mistakes when they try to use rafter ties as ceiling joists in semivaulted ceilings. For maximum headroom or aesthetic balance, they place the rafter ties halfway up the roof pitch, near the center of the rafter span, where they're too high to be an effective tie. Applying the insulation and the drywall greatly increases the load on the rafters at their most critical point: midspan (what engineers call the maximum bending moment). This added load can cause the rafters to sag, pulling the ridge down and also pushing the exterior walls outward.

To avoid this problem, you'd need to engineer the rafters to carry the point load created by the additional weight being placed on them. You'd also need to design a ridge beam capable of supporting

THE WRONG AND THE RIGHT OF RAFTER TIES

To prevent roof loads from spreading the walls outward, rafter ties (or ceiling joists) must be in the lower third of the roof pitch. Collar ties are too high to keep walls from spreading and instead serve to resist uplift by holding the rafters together at the ridge.

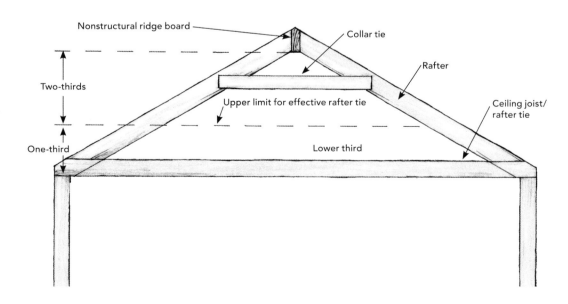

Nonstructural ridge board

Collar tie

Rafter

Two-thirds

Upper limit for effective rafter tie

Ceiling joist/ rafter tie

One-third

Lower third

the roof load, just as you would if it were a cathedral ceiling, which essentially it is.

Trusses Require Precise Permanent Bracing

Any builder who's ever heard the words domino effect knows it's important to brace trusses as they are being erected. But not everybody understands what permanent bracing involves.

To ensure a stable, long-lasting roof, most truss systems require three types of permanent bracing: continuous lateral bracing of the top and the bottom chords and diagonal bracing at the end of the building, and in between if necessary. The bracing for the top chord is typically satisfied when the roof sheathing is applied. The bottom-chord bracing is normally accomplished by placing a row of 2×4s on top of the bottom chord and then running them alongside a panel point (the point where the webs

and the bottom chord meet) for the full length of the building. In a wide building, these bottom-chord braces should be roughly 10 ft. on center.

The diagonal bracing—actually a form of X-bracing—is the one builders often get wrong or omit altogether. Diagonal bracing should be placed at each end of the building and every 25 ft. along the length of a long building. To prevent the domino effect, the first leg of the X is formed by a 16-ft. 2×4 running down at a 45-degree angle (or less) from the ridgeline of the gable-end truss to the bottom chord of the farthest reachable inner truss. To take the hinge effect out of the connection between the gable wall and the gable truss, the other leg of the X runs from the top plate of the gable-end wall upward to the top chord of the same inner truss to which the first leg is attached. It's also important to make sure the braces are run alongside the webs of the intervening trusses and securely nailed to each truss.

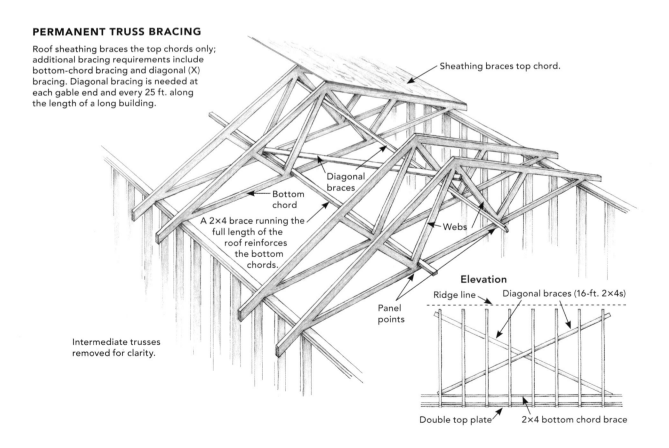

PERMANENT TRUSS BRACING

Roof sheathing braces the top chords only; additional bracing requirements include bottom-chord bracing and diagonal (X) bracing. Diagonal bracing is needed at each gable end and every 25 ft. along the length of a long building.

Sheathing braces top chord.

Diagonal braces

Bottom chord

A 2×4 brace running the full length of the roof reinforces the bottom chords.

Webs

Intermediate trusses removed for clarity.

Panel points

Elevation

Ridge line

Diagonal braces (16-ft. 2×4s)

Double top plate

2×4 bottom chord brace

In certain situations, it may also be necessary to brace long web members that are in compression to prevent them from buckling under load. If any web bracing is required, the proper procedure for it will be noted in the technical design sheet that comes from the manufacturer.

Loads Must Be Placed Carefully atop Trusses

After the trusses are placed and braced—and before the crane operator is allowed to leave—many builders lift pallet loads of sheathing panels onto the trusses for easy access by the framers. Although this practice may be convenient, it can greatly overload the trusses if not done carefully.

On steep roofs, the crew typically erects a temporary platform to hold sheathing. This platform usually consists of two legs resting on the trusses, plus some framework to support the sheathing horizontally (see the sidebar below). This arrangement puts most of the load on two trusses. The higher on the roof these loads are placed, the greater the stress factor. Severe bowing or total collapse of the trusses can result if the stress is too great.

The proper way to load pallets of sheathing on the roof is to place the load as low as possible on the trusses. The type of platform I just described is fine, but set the legs down onto the top plate of the exterior wall. This way, the wall—not the trusses—shoulders most of the burden.

WHICH ROOF WOULD YOU RATHER SHEATHE?

IN THE PHOTO BELOW, THE SHEATHING HAS BEEN stacked too high on this steep roof and is seriously overloading several trusses. A strong gust of wind is all it may take to bring down this house of cards. In the photo at right, even though it's more than a full pallet, the load is low on the roof, and the weight is bearing on the walls.

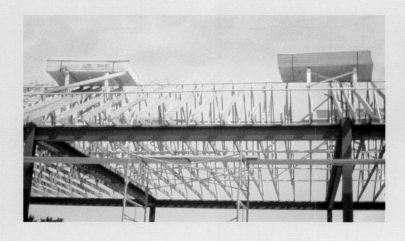

Wind-Resistant Framing Techniques

BY BRYAN READLING

You've seen photos and videos of massive tornadoes ripping through towns and wiping neighborhoods off the map. Given the destruction, you might guess that any house close to an advancing tornado is doomed. The reality, though, is that weaker twisters—those rated EF-0, EF-1, and EF-2 by the National Weather Service—make up 95 percent of all tornadoes. A carefully constructed house often can survive a hit from one of these smaller, more common storms.

As an engineer for APA–The Engineered Wood Association, I spend a lot of my time studying wind damage to houses and figuring out ways to boost a house's resistance to hurricanes, tornadoes, and windstorms. My work includes plenty of travel, because tornadoes and hurricanes affect most of the country and high-wind events happen everywhere.

My latest field-research project was in April 2011, when two storms two weeks apart spawned tornadoes in seven Southern states. The second storm caused the single largest tornado outbreak in recorded history. In our subsequent investigation of wind-damaged houses 10 years old and newer, my colleagues and I discovered that most of the structural failures were caused by a lack of continuity in the load path that connects a house's structural elements from the foundation to the roof.

SEEING STORM DAMAGE FIRSTHAND. Following an outbreak of deadly tornadoes in the spring of 2011, the author surveyed hundreds of storm-damaged homes in a multistate investigation. His research revealed that many of the most-badly damaged homes could have fared much better had their builders adopted a few simple and inexpensive wind-resistant framing methods.

MISSING ANCHOR BOLTS. This house, which was attached to its masonry foundation with cut nails, was pushed 6 ft. off its foundation by tornado-driven wind. Similar failures occurred with houses that were nailed to slab foundations.

UNBACKED FOAM SHEATHING. Foam sheathing performs better when the interior is covered with drywall. Gable ends without drywall, like the gables on these neighboring homes, should be sheathed with structural panels.

MISSING CONNECTORS. Unless it's adequately secured, roof framing can be pulled from the walls that it's attached to during high winds. Toenailed roof framing ripped from walls was the most commonly observed serious building failure in the author's post-storm research.

CONNECTORS ON THE WRONG SIDE. This house's roof framing was attached to the walls' top plate with metal hurricane ties. Unfortunately, the ties were fastened on the inside of the top plate, where they aren't as strong as connectors aligned with the wall sheathing on the exterior.

Roof Failures

The most common and often most devastating load-path failures occurred when rafters and trusses were pulled from exterior walls. Many of the most severely damaged houses had roof framing attached to the walls with toenails, an inherently weak connection because it relies on the nails' withdrawal capacity. Modern building codes allow toenailed rafters in most non-hurricane areas, but many engineers don't believe toenails have the strength to meet some *International Residential Code* requirements.

Roof failures were not limited to houses with toenailed trusses and rafters. Failures also occurred when metal hurricane ties were nailed on the interior of the top plate instead of the exterior. Exterior-mounted metal connectors hold better because they line up with the wall sheathing's load path.

BEYOND CODE FOR HIGH-WIND RESISTANCE

THE FRAMING DETAILS SHOWN HERE ARE NOT COMPLICATED or expensive to execute when they are incorporated into the plans for a new house. In addition to these measures, there are other ways to protect houses in hurricane- and tornado-prone areas.

First, protect large openings. Picture windows, sliding-glass doors, garage doors, and other large openings are vulnerable to damage in high-wind events. Breaches can lead to pressurization of the building interior and increased loads on the structure. Consider installing windows, doors, and garage doors rated for high winds and impact damage.

While a stronger, more wind-resistant structure is certainly safer for occupants, think about adding a safe room in a basement or central space.

Finally, consider using hip roofs, which are more aerodynamic and provide better support to the tops of exterior walls than gable roofs.

Tie down rafters

Secure rafters and trusses with metal connectors. The roof-to-wall connection is subject to both uplift and shear. Inexpensive framing connectors make this important connection simple. Place connectors on the outside of the wall, where they'll do the most good.

Use enough nails

Nail wall sheathing with 8d common (0.131 in. by 2½ in.) nails 4 in. on center at ends and edges and 6 in. on center in the intermediate framing. This installation will greatly increase wind and racking resistance compared to code-minimum requirements.

Lap the sill

Extend wood structural-panel sheathing to the sill plate. The connection of the wall sheathing to a properly anchored sill plate is an important part of the load path. Available at many pro-oriented lumberyards, 9-ft.-long and 10-ft.-long OSB simplifies this connection.

Bolt sill plates

Anchor sill plates with ½-in. anchor bolts equipped with 0.229-in.-thick, 3-in. by 3-in. square plate washers. Space the bolts from 32 in. to 48 in. on center. The IRC requires a minimum spacing of 6 ft. for houses subjected to wind speeds up to 110 mph, but tighter spacing greatly improves wall performance.

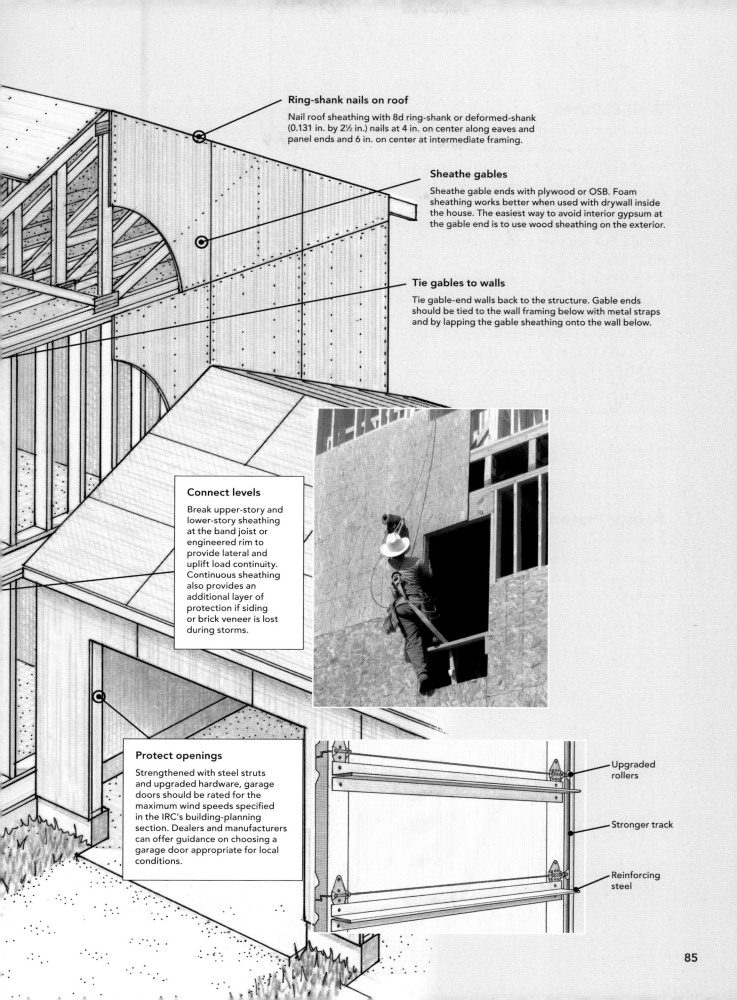

Ring-shank nails on roof

Nail roof sheathing with 8d ring-shank or deformed-shank (0.131 in. by 2½ in.) nails at 4 in. on center along eaves and panel ends and 6 in. on center at intermediate framing.

Sheathe gables

Sheathe gable ends with plywood or OSB. Foam sheathing works better when used with drywall inside the house. The easiest way to avoid interior gypsum at the gable end is to use wood sheathing on the exterior.

Tie gables to walls

Tie gable-end walls back to the structure. Gable ends should be tied to the wall framing below with metal straps and by lapping the gable sheathing onto the wall below.

Connect levels

Break upper-story and lower-story sheathing at the band joist or engineered rim to provide lateral and uplift load continuity. Continuous sheathing also provides an additional layer of protection if siding or brick veneer is lost during storms.

Protect openings

Strengthened with steel struts and upgraded hardware, garage doors should be rated for the maximum wind speeds specified in the IRC's building-planning section. Dealers and manufacturers can offer guidance on choosing a garage door appropriate for local conditions.

Upgraded rollers

Stronger track

Reinforcing steel

Wall Failures

Another common observation, especially in the hardest-hit areas, was houses blown off their foundations. Most had their walls attached to the foundation with hand-driven, cut masonry nails and, in a few locations, pneumatic framing nails. Obviously, anchor bolts are a better choice, especially when the bolts have large square washers to prevent them from pulling through the plate.

A gable end is often poorly connected to the rest of the building. We saw many houses where the triangle-shaped gable end had blown in, often leading to greater damage from wind and water. The gable end is especially vulnerable to failure because its walls are often not backed with drywall. Walls backed with drywall in living space generally hold up better because the drywall provides additional resistance to wind and debris. Failures like this were even more common when the gables were covered with foam sheathing and vinyl siding because both materials are vulnerable to wind pressure and flying debris.

Poorly Fastened Sheathing

When the houses we studied were at least partially intact, the loss of wood wall and roof sheathing often could be attributed to improper attachment. Nails used as prescribed in the building codes provided good performance, whereas staples performed poorly because they offer less pullout resistance than nails and must be used in greater quantity. Poorly attached roof sheathing at the last rafter or gable-end truss was identified as a weak link in roof construction.

We also saw many cases where breaches in the exterior walls due to wind pressure or flying debris caused pressurization of the building, sometimes resulting in homes that blew apart completely. Field and wind-tunnel research has revealed that wind and flying-debris damage to doors, windows, and nonstructural claddings like brick and vinyl siding often lead to more catastrophic structural failures. Large openings such as garage doors are especially vulnerable to impact and wind-pressure damage.

A Small Price to Pay

Most of these above-code improvements are easy to implement and surprisingly affordable. In researching the 2013 Georgia Disaster Resilient Building Code, the Georgia Department of Community Affairs determined that the added cost of implementing the APA's recommendations is about $595. This estimate, which includes materials and labor, is based on a 2,100-sq.-ft. slab-on-grade ranch house with a 10-in-12 roof pitch and three gables.

Where Do You Want the Blocking?

BY JUSTIN FINK

The final 5 percent of any good framing job is blocking. It makes work easier for the subsequent tradesmen and future homeowners, and it can be completed using cutoffs that would otherwise land in the Dumpster. So why doesn't every house have sturdy blocking behind towel bars, under stairs, and in closets? Some of it is eliminated to save time and money, and some is overlooked. It's also a good bet that lots of blocking is left out simply because nobody ever asked the right people where it was needed. But what if the plumber, the electrician, the drywall contractor, and the finish carpenter showed up on the job before the framers rode off into the sunset?

To explore this scenario, I asked five of our frequent contributors to help create a blocking wish list. Every house is different, of course, and this list isn't complete. It does, however, provide a useful road map to a desirable destination: solid backing for many of the fixtures, appliances, trim details, and other common features found in a typical house.

Every Trade Depends on Solid Blocking

FIVE TRADESMEN WEIGH IN ON WHERE TO put solid blocking for stair skirts, grab-bar anchors, and everything else the code leaves out.

GARY M. KATZ
Finish carpentry

MIKE GUERTIN
Framing and
general carpentry

ED CUNHA
Plumbing

CLIFFORD A. POPEJOY
Electrical

MYRON R. FERGUSON
Drywall

KITCHEN BLOCKING

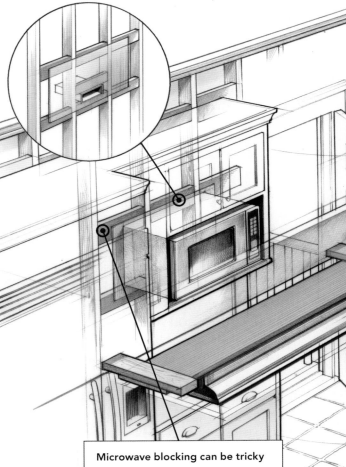

Microwave blocking can be tricky

Microwaves have a mounting panel that needs to be secured to the wall. Although the panel can be attached to a stud, I feel more confident when it's screwed to blocking. If the microwave is over a cooktop, the key is to locate the blocking to avoid the area where exhaust-venting ductwork runs through the stud cavity. If the venting runs down through the wall, I put a block above the opening. If it runs up through the wall, I put a block below the opening. If it's direct vent (as shown in the inset drawing), I block above and below the opening. —M.G.

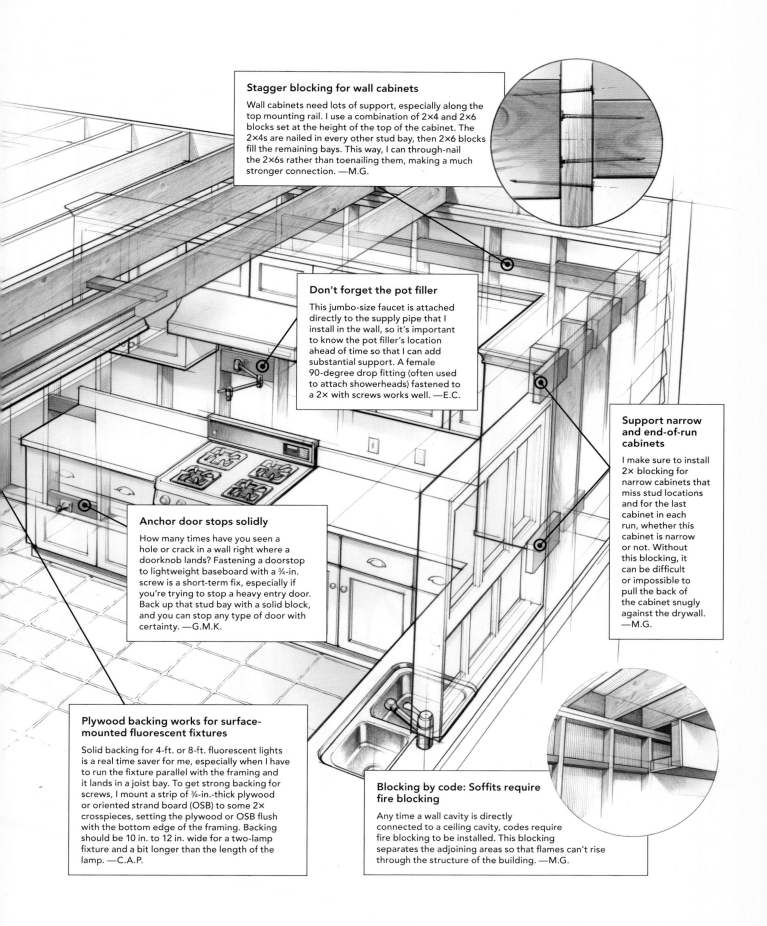

Stagger blocking for wall cabinets

Wall cabinets need lots of support, especially along the top mounting rail. I use a combination of 2×4 and 2×6 blocks set at the height of the top of the cabinet. The 2×4s are nailed in every other stud bay, then 2×6 blocks fill the remaining bays. This way, I can through-nail the 2×6s rather than toenailing them, making a much stronger connection. —M.G.

Don't forget the pot filler

This jumbo-size faucet is attached directly to the supply pipe that I install in the wall, so it's important to know the pot filler's location ahead of time so that I can add substantial support. A female 90-degree drop fitting (often used to attach showerheads) fastened to a 2× with screws works well. —E.C.

Support narrow and end-of-run cabinets

I make sure to install 2× blocking for narrow cabinets that miss stud locations and for the last cabinet in each run, whether this cabinet is narrow or not. Without this blocking, it can be difficult or impossible to pull the back of the cabinet snugly against the drywall. —M.G.

Anchor door stops solidly

How many times have you seen a hole or crack in a wall right where a doorknob lands? Fastening a doorstop to lightweight baseboard with a ¾-in. screw is a short-term fix, especially if you're trying to stop a heavy entry door. Back up that stud bay with a solid block, and you can stop any type of door with certainty. —G.M.K.

Plywood backing works for surface-mounted fluorescent fixtures

Solid backing for 4-ft. or 8-ft. fluorescent lights is a real time saver for me, especially when I have to run the fixture parallel with the framing and it lands in a joist bay. To get strong backing for screws, I mount a strip of ¾-in.-thick plywood or oriented strand board (OSB) to some 2× crosspieces, setting the plywood or OSB flush with the bottom edge of the framing. Backing should be 10 in. to 12 in. wide for a two-lamp fixture and a bit longer than the length of the lamp. —C.A.P.

Blocking by code: Soffits require fire blocking

Any time a wall cavity is directly connected to a ceiling cavity, codes require fire blocking to be installed. This blocking separates the adjoining areas so that flames can't rise through the structure of the building. —M.G.

BATHROOM BLOCKING

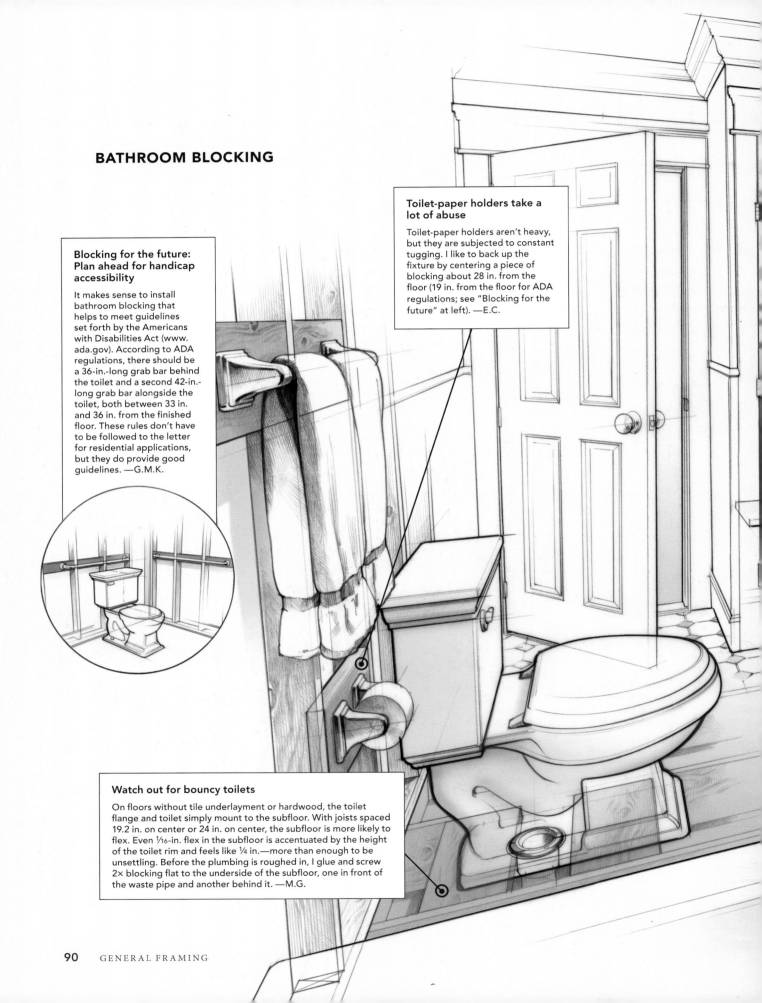

Blocking for the future: Plan ahead for handicap accessibility

It makes sense to install bathroom blocking that helps to meet guidelines set forth by the Americans with Disabilities Act (www.ada.gov). According to ADA regulations, there should be a 36-in.-long grab bar behind the toilet and a second 42-in.-long grab bar alongside the toilet, both between 33 in. and 36 in. from the finished floor. These rules don't have to be followed to the letter for residential applications, but they do provide good guidelines. —G.M.K.

Toilet-paper holders take a lot of abuse

Toilet-paper holders aren't heavy, but they are subjected to constant tugging. I like to back up the fixture by centering a piece of blocking about 28 in. from the floor (19 in. from the floor for ADA regulations; see "Blocking for the future" at left). —E.C.

Watch out for bouncy toilets

On floors without tile underlayment or hardwood, the toilet flange and toilet simply mount to the subfloor. With joists spaced 19.2 in. on center or 24 in. on center, the subfloor is more likely to flex. Even 1/16-in. flex in the subfloor is accentuated by the height of the toilet rim and feels like 1/4 in.—more than enough to be unsettling. Before the plumbing is roughed in, I glue and screw 2× blocking flat to the underside of the subfloor, one in front of the waste pipe and another behind it. —M.G.

Standard closets get a standard treatment

Unless the homeowner has a specific shelving layout in mind, I reinforce all the usual trouble spots in a closet. For a standard 2-ft.-deep closet, this means 2×4s on the flat about 12½ in. out from the back wall to support closet-rod cups, and sister blocks on the back-wall corner studs, starting 18 in. off the floor and extending to about 1 ft. below the ceiling, to help support shelving. —M.G.

Install light fixtures right where they belong

I like to place the box for a light fixture in the right spot, not just on the closest ceiling joist. To do this, I fasten the electrical box to a 2×4 block, then nail or screw that block between the joists. I do the same for wall-mounted fixtures like sconces: Nail on the box, and toenail the block between the studs. —C.A.P.

Plan for pedestal sinks

Blocking for a pedestal sink isn't just for convenience. I need to anchor the bolts on the back of the sink to something solid to keep the pedestal upright. Installing blocking after the drywall is hung is difficult; if the walls are tiled, it's near impossible. If you plan to have a pedestal sink, put a 2×8 between the studs behind the sink, centered at about 26 in. above the height of the finished floor. Just in case the fixture specs change, there is enough meat to accommodate most pedestals. —E.C.

Towel racks need proper support

Towel racks are often installed with hollow-wall anchors, and that's why they often fall off the wall. Wet towels are heavy, and these bars take abuse; use solid 2× blocking. —E.C.

Provide support for horizontal electrical receptacles

In some cases, like a bathroom backsplash with a mirror above, there might not be enough height for a standard switch plate in the traditional vertical orientation. Laying the box on its side is often the best solution. Most single-gang electrical boxes are set for side-nailing, so I just toe-screw or toenail a 2× on the flat between the studs and fasten the box to the blocking as if it were a normal vertical stud. —C.A.P.

Use vertical backing for the tub surround

An extra vertical 2× is a good idea for most tub surrounds, or tubs that will have tile backer installed above. Locating the blocking 30 in. out from the corner will work for average tubs, but check your plans to be sure. —M.G.

LIVING ROOM BLOCKING

Big crown molding needs wide support

I've installed miles of traditional crown molding by cross-nailing into the ceiling in spots where I couldn't find a joist, and I've never had a joint fail. On some jobs, though, the built-up crown molding can project nearly a foot across the ceiling. In these cases, having solid 2×4 or 2×6 blocking in all joist bays helps to speed up the finish work enormously. —G.M.K.

Blocking by code: Keep fire blocking high and flush

Codes require fire blocking in walls that are 10 ft. tall or more. But when the blocking is installed 4 ft. or 8 ft. off the floor, it lands right behind the long seams of drywall and causes the joint to ridge out. Instead, I like this blocking set at 6 ft. so that it lands in the middle of a sheet of drywall. If the blocking is toenailed, make sure the nails are set flush. —M.R.F.

Back up low-voltage accessories that don't use boxes

Security-system hardware, doorbell chimes, and other low-voltage electrical accessories don't mount in electrical boxes. To install these fixtures properly without relying on hollow-wall anchors, I toenail in a piece of 2×4 blocking. Also, I drill a ¼-in. hole in the block to route the cable, then wrap it around a nail or screw and secure it with electrical tape so that the drywall contractors are less likely to cover it up. —C.A.P.

Plan for vertical wainscoting

Lots of guys like to install tongue-and-groove wainscot on top of plywood or OSB, but because of fire codes in my area, I install it over ⅝-in. drywall—a poor substitute for blocking. I also like to sit the wainscot on top of a solid backerboard an inch or two above the baseboard, which means I can't nail the wainscoting into the bottom plate. So, I like a row of blocks at the bottom of the wall and a second row where the top of the wainscot and the chair rail will land. —G.M.K.

Fur out switch boxes near trimmed openings

Light-switch boxes are usually placed at the door, but attaching the box to the king stud that's part of the rough opening might put the switch plate in the space to be occupied by the door casing. Sure, I could put the box on the far side of the stud bay, but I more often use a couple of long 2× blocks set about 48 in. above the floor to space the box clear of the trim zone. —C.A.P.

Fireplace surrounds are often forgotten

Anyone who has ever installed a mantel knows there's never any more than a stud or two on either side of the fireplace; that's a real problem when it comes to attaching the wide pilasters for a fireplace surround. Although it isn't that tough to install plywood or OSB on top of drywall with plastic plugs and adhesive caulking, having solid horizontal blocking sure makes the job easier. —G.M.K.

Corner blocking should be wider

The stud configurations of inside corners vary, but almost every house in which I've installed drywall has 2×4 blocking for 2×4 stud walls. The trouble is that this leaves only 1 in. to 1½ in. of exposed surface to fasten to, depending on which wall is covered first, and I'm forced to wedge the nose of a screw gun into the corner and drive the fastener at an angle. If I had my way, framers would swap the normal 2×4 for a 2×6, then install a second piece of blocking along the other wall. —M.R.F.

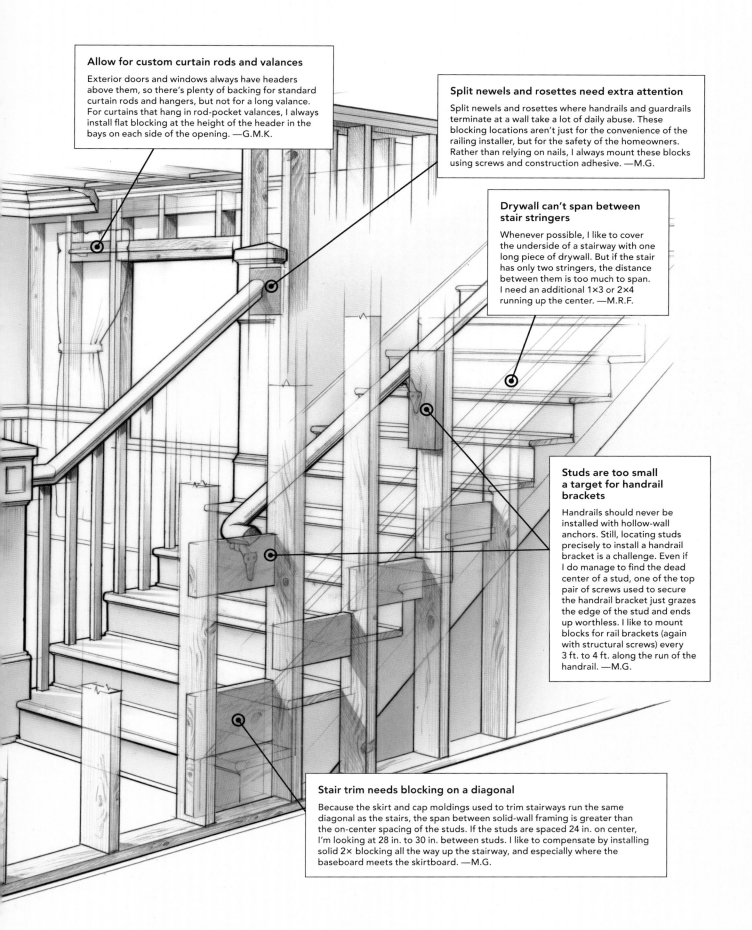

Allow for custom curtain rods and valances

Exterior doors and windows always have headers above them, so there's plenty of backing for standard curtain rods and hangers, but not for a long valance. For curtains that hang in rod-pocket valances, I always install flat blocking at the height of the header in the bays on each side of the opening. —G.M.K.

Split newels and rosettes need extra attention

Split newels and rosettes where handrails and guardrails terminate at a wall take a lot of daily abuse. These blocking locations aren't just for the convenience of the railing installer, but for the safety of the homeowners. Rather than relying on nails, I always mount these blocks using screws and construction adhesive. —M.G.

Drywall can't span between stair stringers

Whenever possible, I like to cover the underside of a stairway with one long piece of drywall. But if the stair has only two stringers, the distance between them is too much to span. I need an additional 1×3 or 2×4 running up the center. —M.R.F.

Studs are too small a target for handrail brackets

Handrails should never be installed with hollow-wall anchors. Still, locating studs precisely to install a handrail bracket is a challenge. Even if I do manage to find the dead center of a stud, one of the top pair of screws used to secure the handrail bracket just grazes the edge of the stud and ends up worthless. I like to mount blocks for rail brackets (again with structural screws) every 3 ft. to 4 ft. along the run of the handrail. —M.G.

Stair trim needs blocking on a diagonal

Because the skirt and cap moldings used to trim stairways run the same diagonal as the stairs, the span between solid-wall framing is greater than the on-center spacing of the studs. If the studs are spaced 24 in. on center, I'm looking at 28 in. to 30 in. between studs. I like to compensate by installing solid 2× blocking all the way up the stairway, and especially where the baseboard meets the skirtboard. —M.G.

Framing Floors

The Well-Framed Floor

BY JIM ANDERSON

Walking across a newly framed floor for the first time is a milestone in any framing project. Finally, there's something to stand on that doesn't squish beneath your boots. It's flat and strong, and because there's a floor to stand on, the rest of the project will move ahead much more quickly. But whether you're using common lumber or I-joists (see the sidebar on p. 96), it takes a well-coordinated effort to get any floor to the point where you can walk on it.

Before you start driving nails, it's important to collect as much information as possible about the locations of the joists, posts, beams, point loads, cantilevers, plumbing vents, drains, and HVAC ducts on the floor-framing plan. Whether those details come from the architect, you, or somewhere else, the floor-framing plan needs to reflect the house as it's going to be built.

PLAN AHEAD. Whether it's the floor of a big house or a small addition, an accurate layout and efficient techniques promote smooth installation.

WHY I PREFER I-JOISTS OVER SOLID WOOD

I REMEMBER THE first time I saw I-joists, those long, floppy things. They seemed so flimsy and light that I thought they would have trouble holding up the sheathing, not to mention the walls that would go on top of them.

They have more than proven me wrong, however. The main advantages are that I-joists are dimensionally stable and very straight. The web (the wide middle section) of an I-joist is cut from oriented strand board, thin strands of wood oriented in the same direction and glued together. Because glue surrounds all those strands of wood, you can expect less shrinking and swelling and very consistent joist sizes (usually within 1/16 in.).

You also can cut much larger holes into I-joists than into solid lumber; holes up to 6 in. are allowed in the center of the span of a 9½-in. I-joist. Elsewhere along the web, 1½-in. holes are provided in perforated knockouts. Holes in solid lumber can be no more than one-third the total width.

I-joists must be handled carefully; upright is best, or supported in a couple of places if carried flat. They're light, come in lengths up to 60 ft., and can span long distances as part of an engineered floor system. Best of all, they cost about the same as lumber; in the longer lengths, they actually cost less.

ADJUST THE LAYOUT BEFORE IT'S TOO LATE

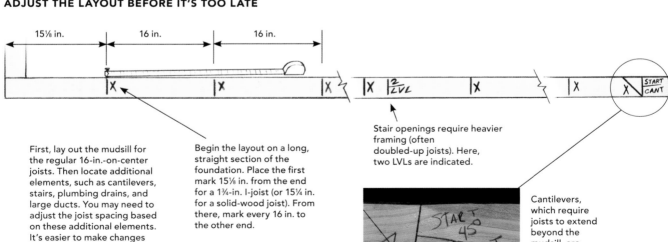

15⅛ in.   16 in.   16 in.

2 LVL

X   START CANT

First, lay out the mudsill for the regular 16-in.-on-center joists. Then locate additional elements, such as cantilevers, stairs, plumbing drains, and large ducts. You may need to adjust the joist spacing based on these additional elements. It's easier to make changes now than later.

Begin the layout on a long, straight section of the foundation. Place the first mark 15⅛ in. from the end for a 1¾-in. I-joist (or 15¼ in. for a solid-wood joist). From there, mark every 16 in. to the other end.

Stair openings require heavier framing (often doubled-up joists). Here, two LVLs are indicated.

START 45 CANT

Cantilevers, which require joists to extend beyond the mudsill, are labeled to indicate their angle and starting point.

Having all this information in one place allows you to overlay—in pencil—the big immovable parts of the house on top of each other. This step will catch most if not all the big mistakes that can be made early on. It's a lot easier to erase than it is to remove and replace.

Transfer the Details from the Plans to the Mudsills

First, I check the joist spacing on the floor-framing plan, usually 16 in. or 19.2 in. on center, and transfer that to the mudsills. Measuring from the end of the house (usually beginning with the longest uninterrupted run), I mark the edge of the first joist 15⅛ in. from the end for 1¾-in. I-joists (16 in. minus half the joist thickness). This places the center of the first joist at 16 in.

Then I mark 16 in. on center (or whatever the proper spacing is) from the first mark to the other end of the house. I do this on the front and back walls, then I check the layout marks on both ends to make sure that they are the same. If they are within ¼ in., I leave them; if not, I double-check the layout and make adjustments. I also mark the location of stairs, load-bearing members, and cantilevers on the mudsill.

Leave Room for Pipes and Ductwork

If the layout mark for the last joist is within a foot of the endwall, I move it to allow room for plumbing, electrical, or HVAC in what is often an important joist bay. I usually just measure and mark 16 in. from the edge of the mudsill back toward the center of the house.

I also make sure that none of the plumbing fixtures or flue chases land on a joist. This is another opportunity to double-check myself. It's a lot easier to move the joist now than it is to move it later or repair damage from a determined plumber with a chainsaw. I usually allow a minimum of 12 in. between joists for furnace flues, which provides 2 in. of clearance on each side for an 8-in. furnace flue. Even though 1 in. on each side meets the building code here in Denver, I figure that where heat and wood are concerned, more room is better.

Again, I create this space either by moving the joist off the 16-in.-on-center layout or, when that isn't practical, by cutting the joist just short of the flue and supporting it with a header tied into the joists on each side of the one that's cut.

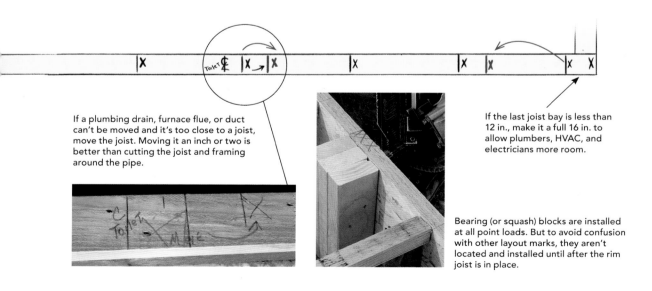

If a plumbing drain, furnace flue, or duct can't be moved and it's too close to a joist, move the joist. Moving it an inch or two is better than cutting the joist and framing around the pipe.

If the last joist bay is less than 12 in., make it a full 16 in. to allow plumbers, HVAC, and electricians more room.

Bearing (or squash) blocks are installed at all point loads. But to avoid confusion with other layout marks, they aren't located and installed until after the rim joist is in place.

SQUARE ONE END OF EACH JOIST WHILE SORTING AND STACKING. Because I-joists are cut to approximate length at the lumberyard, it is easier just to square one end as you are sorting them.

Plumbing drains and supply lines are zero-clearance items, so I can have wood right next to them. I locate the fixtures on the plan, and if a joist is on or near the centerline of the drain, I move the joist 1 in. or 2 in. in one direction or the other. If I have two fixtures close together and moving a joist away from one drain places it beneath another, I open the spacing a little more (and double the joist) so that both drains lie within a slightly oversize bay.

Prepare Material According to Where It's Needed

Wood I-joists come from the yard in a large bundle; the rim material and any LVLs usually are strapped to the top. With a helper, I move the LVLs to sawhorses for cutting to length and to install joist hangers.

We move the rim joists to the top of the sheathing or to the ground, and place stickers beneath so that we can lift them easily later. Then we square one end of all the wood I-joists with a simple jig as we take them off the pile and sort them by length and location. When I finish with the I-joists, I build any LVL headers and add joist hangers if they're needed.

After the prep work is done, I usually call in a crane to set all the steel beams that will carry the first floor and to spread all the presorted stacks of joists and LVLs to their appropriate locations. I also move the sheathing to within 3 ft. or 4 ft. of the foundation so that I don't have to carry it any farther than necessary.

After placing the steel, I make sure that the layout on the beams matches what is on the walls. I check the layout by pulling a string from front to back to verify that the layout marks on the front and back walls intersect the marks on the beams. I also make sure that the beams are straight and flat, and make any necessary adjustments.

Spread Joists to the Layout Marks and Roll Them Upright

With one person on each side of the foundation, we quickly position the joists on their layout marks, with the square-cut end aligned on the rim-joist line snapped along the mudsill. Then we tack each joist in place with an 8d nail to keep it in place. It's easier to set the joists to the line first and then install the rim joist later. After tacking down all the joists, we prepare to cut the other end of the joists in place. We snap a chalkline that is 1⅜ in. from the outside of the mudsill, which becomes the cutline.

Cutting the joists to their finished length is as simple as running the saw along the chalkline using the I-joist cutoff guide. The scrap of wood lands in front, where it's available for use as a piece of blocking.

We position one person in the front and one in the back, and starting from one end, we stand all the joists and nail them in place. The 8d nail that had held the joist in place now acts as a hinge for it. We usually can stand all of the joists for 40 lin. ft. of floor in about 10 minutes.

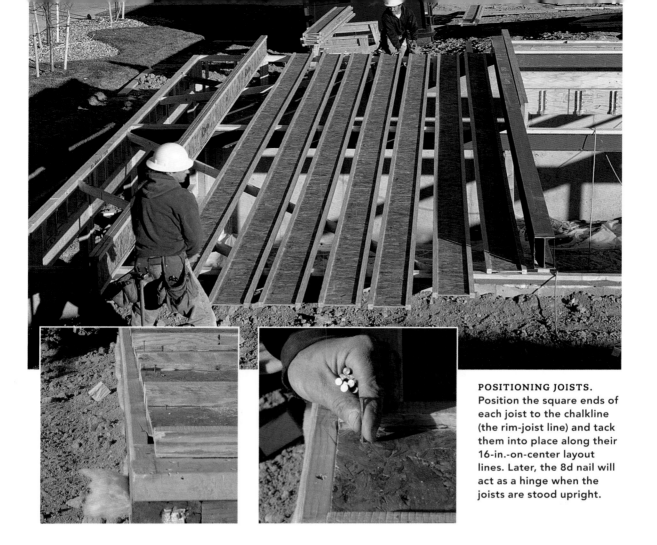

POSITIONING JOISTS.
Position the square ends of
each joist to the chalkline
(the rim-joist line) and tack
them into place along their
16-in.-on-center layout
lines. Later, the 8d nail will
act as a hinge when the
joists are stood upright.

On each end of the I-joist and at the center beam, we put one 10d nail on each side of the joist through the flange into the mudsill. We keep the nails as far from the end of the joist as possible to avoid splitting the I-joist's flange. After standing the joists, we add the rim boards, cutting and nailing as we work our way around the house. We put one 10d nail through the rim into the top and bottom flange of each I-joist.

Once the rim joist goes up, the last thing to do before sheathing is to add bearing blocks, also known as squash blocks. One person details the rim joist for bearing blocks, and another follows behind and nails them in place.

Bearing blocks are required anywhere that concentrated loads land on the joists, such as doorways or where a post supports a beam. We also put them at all inside corners, because 90 percent of the time this spot is a bearing point.

Stack Sheathing on the Floor as Soon as Possible

We snap the line for the first course of sheathing 48½ in. from the outside edge of the rim joist. It's held back a little from the rim to account for any inconsistency in the rim joist.

Before we begin nailing the sheathing, we look for joists that may have been moved from the 16-in.-on-center layout. If our plywood joints are able to avoid them, sheathing will go much faster. After deciding on a starting point, we spread construction adhesive on top of the joists. Then we lay the first row and two sheets of the second row. This approach creates a little staging area where we can stack the rest of the sheathing.

We sheathe over to the steel beams in the center and add any bearing blocks and joist blocking when we get there. Waiting until the floor is partially

sheathed before installing blocks is a lot easier and safer than trying to balance on unbraced joists.

We cut all the blocks and spread them across the edge of the sheathing (next to the beam), starting at one end and grabbing them off the sheathing as we go. Layout marks for each joist on the plywood's edge keep the joists straight and plumb, and the spacing for blocking consistent. When we have a finished basement, we also add wall ties as we work our way across the floor, which keeps us from having to walk across unsupported joists.

YOU CAN DO THIS ALONE, BUT IT SURE GOES QUICKER WITH TWO. With one person at each end, stand the joists upright and put them on their layout marks. Drive one 10d nail through the flange on each side of the joist into the mudsill (or pony wall).

AFTER ALIGNING THE JOISTS, snap a chalkline and cut them to length in place. Beware of anchor bolts lurking below when making this cut.

THE RIM JOIST GOES ON AFTER THE JOISTS ARE IN PLACE. The rim joists are cut and nailed to the mudsill every 8 in. with a 10d nail.

LAY THE FIRST ROW OF SHEATHING PLUS TWO MORE SHEETS. Then move the rest of the stack onto the floor. It takes about 10 minutes to move 40 sheets; it's much quicker than having to climb up and down to get every sheet.

SHEATHE YOUR WAY OVER TO WHERE BLOCKING IS NEEDED. Do not walk across unstable joists or work from a ladder below the floor. Sheathe over to the beam, then add the joist blocking.

CUT THE SHEATHING IN PLACE. Run sheets long at the ends, snap a chalkline, and cut off the excess. This process is faster and turns out a better floor than cutting each piece to fit.

When we get to a stair rough opening, we sheathe over it and brace the plywood seams. Not only is this approach safer, it also creates more usable floor space when we start framing walls. Before we stand any walls that surround the stair opening, we open it up again. If the hole is too large to sheathe over, we add a safety rail.

Lay as Many Full Sheets as Possible

As we sheathe, we lay as many full sheets as possible (making the fewest number of cuts). I've found that running the sheets long at the ends and cutting to a chalkline snapped along the rim turns out a better product than measuring and cutting the pieces to fit individually.

I pull the chalkline in an extra ⅛ in. from the outside of the rim; this eliminates ever having to cut the rim line again. One person starts at a corner of the house and snaps all the rim lines; the other follows behind with the saw. The rim joist is first straightened and then nailed to the sheathing every 6 in.

Fast, Accurate Floor Sheathing

BY DANNY KELLY

As a carpenter turned general contractor, I'm always happy when we start installing the subfloor on the first level of a new house. Floor sheathing means that we can finally stop slogging around in the mud and will soon have a nice level surface for setting up tools and ladders. Sheathing a floor like the 1,800-sq.-ft. one shown here can take all day with an inexperienced crew, but the guys I work with were able to bang out this floor in a little under an hour.

On most houses, the longest exterior wall perpendicular to the floor joists is the place to start sheath-ing a floor. With a four-person crew, two carpenters move and cut panels, and two place and nail the sheets to the joists.

We cut the sheets to length after they've been nailed in place; this is faster and eliminates layout and measuring mistakes. We're careful to cut floor sheathing flush with the band joist. Otherwise, over-hanging pieces will prevent the wall sheathing from fitting tight to the band joist. To get the sheets to fall on the center of the floor joists, subtract ¾ in. from the first joist cavity. Then the first sheet can start flush with the band joist without any waste.

LAY DOWN A WORK SURFACE

IT'S TOO DANGEROUS TO WALK ON THE TOPS OF joists while sheathing a subfloor, so create a work surface by covering the floor joists with as many sheets of tongue-and-groove sheathing as you need to move around safely. Arrange the panels so that all the tongues and all the grooves are oriented consistently. Start with a full sheet on the corner of the longest exterior wall perpendicular to the joists. The next row starts with a half-sheet. One pair of carpenters should keep laying and nailing down full sheets

while another pair stocks the floor with full sheets, and cuts and places partial sheets. Alternate full sheets and half-sheets to start each row (left). Some sheathing has marks indicating common on-center spacings (right).

HOW TO PUT DOWN A SUB-FLOOR FAST. Two carpenters move panels and make cuts while two position, glue, and nail the sheets. While one sheet is being nailed, another is dragged into place.

LINE IT UP. Line up the edge of one sheet with the previous sheet, making sure that all the tongues and all the grooves are facing the same direction. (The author's crew starts with a tongue toward the outside wall.) Don't drag the panels through the adhesive.

LET IT GO. Drop the panel onto the joists as close to its final position as possible; otherwise, you'll mess up the subfloor adhesive when you slide the panel into place. It doesn't matter whether you drop the tongue or the groove edge.

16 in. on center
19.2 in. on center
24 in. on center

TACK A CORNER. One carpenter moves the sheet so that its leading edge is lined up with the adjacent sheet. Then the nail-gun operator tacks the corner with a single nail. He pauses while his teammate moves the other end into position.

NAIL IT OFF. With the sheet in position, the nail-gun operator drives a nail or two to lock the sheet in its final position and then nails the rest of the sheet. With an experienced team, positioning and tacking take seconds. Panels overhanging the edge of the band joist will be cut in place later.

RUN WILD! Save time and eliminate layout and measuring mistakes by cutting odd-shaped panels in place. The first piece in a row will be a full sheet or a half-sheet, but the last piece at the opposite end likely will be an odd-width offcut.

DO USE A SLEDGE SPARINGLY. The tongues and grooves on subflooring are designed to gap panels properly, so the panels shouldn't be beaten together except when the tongue has been damaged by rough handling. When that's the case, use a sledgehammer to get the sheets to meet up. A board prevents the hammer face from doing additional damage.

DON'T MESS UP THE GLUE. To prevent smearing the subfloor adhesive, stand the sheet on edge in the proper spot, and let it drop into the glue. A well-timed pull with the ball of your foot can help to keep the panel edge close to the previous row.

DO CHECK THE JOIST SPACING. Warped joists don't necessarily line up with on-center spacing, so check the spacing before nailing, then use a hammer to coax joists into the proper position. Subflooring with spacing marks saves time.

MEASURE FROM UNDERNEATH; SNAP LINES ON TOP. Feed a tape measure under the panel until it hits the band joist. Transfer the measurement to the top of the panel, and snap chalklines to guide a circular saw. Set the blade so that it just cuts through the subfloor and doesn't damage the band joist.

PUT OFF CUTTING WHEN YOU CAN. When possible, save the job of cutting unusually sized panels like the ones around this crawlspace opening until you've finished laying all the sheets. Cutting odd-shaped panels is a good job for less experienced carpenters while the more senior carpenters move on to snapping lines and laying out plates for wall framing.

DON'T LET THE GLUE DRY. Apply only as much glue as can be covered with sheathing quickly. Save time by cutting the plastic nozzles on the glue tubes all at once. (An inner seal keeps the opened tubes from drying.) Water-based adhesives, which are more environmentally friendly, work better than they used to, but solvent-based adhesives are still more forgiving in wet weather.

DO USE ENOUGH NAILS. Nail subfloor panels every 12 in. in the field and every 6 in. along panel edges. Keep fasteners ⅜ in. from panel edges for maximum hold. For ¾-in.-thick sheets, use 8d (2½ in.) common nails or gun nails approved by local code. The author uses ring-shank nails and adds screws once the house is dried in.

THE FLOOR ISN'T SQUARE. NOW WHAT?

MY CREW AND I RECENTLY REFRAMED the interior of a 100-year-old brownstone building on Beacon Hill in Boston. The building was so out of square that it was like building in a carnival funhouse. We found that as long as the joists were parallel with one another, installing the subfloor could proceed as normal.

We snapped a line perpendicular to the joists 48 in. from the band joist. The first and last rows were tapered rips, and the first and last pieces in a row were angled. All the field pieces were full, uncut sheets.

—Brian McCarthy is the owner of McCarthy General Contracting in Stow, Massachusetts.

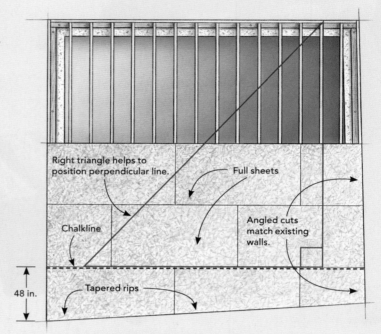

Right triangle helps to position perpendicular line.

Full sheets

Chalkline

Angled cuts match existing walls.

48 in.

Tapered rips

A right triangle is used to create a reference line perpendicular to the floor joists. The line should be 48 in. wide at its widest point to match the panel width.

Built-Up Center Beams

BY RICK ARNOLD AND MIKE GUERTIN

We looked at a basement remodel recently. But before we got to the basement, the owner was showing us large cracks in the tile floor in the kitchen and entry. She hadn't noticed the drywall cracks forming in some doorways. We'd seen these symptoms before, and in the basement we found that the center beam had been built of green lumber and had shrunk almost ¾ in. So before the remodel could begin, we had to jack up that center beam so that it could do its job: Hold up the house.

Built-Up Beam Basics

In the simplest terms, a built-up center beam provides a straight, level surface that supports the floor joists between the walls of the foundation. Like most, the beam we installed for this project was even with the mudsills and was carried by columns set on footings at regular intervals.

A built-up beam is made of several layers of lumber nailed together and set on edge. The beam for this project was made of dimensional lumber, but laminated veneer lumber can also be used. The number of layers and the size of the lumber are determined by the load that the beam has to carry, the species of lumber, and the span between support columns.

KEEP BEAM FLAT DURING PREASSEMBLY. For a straight beam, sections of the beam are preassembled on the ground and must be kept flat, with the tops of the boards kept flush. Just a few nails join the layers at this point.

THE LVL ALTERNATIVE

ALTHOUGH WE INSTALL A LOT OF DIMENSIONAL-lumber center beams, laminated veneer lumber may be a better choice when you want to reduce the number of support columns and simplify the installation. LVL boards span greater distances than similar-size dimensional lumber and are much less prone to shrinking. They also come in longer lengths.

LVL boards are 1¾ in. thick and from 7 in. to more than 18 in. deep. Layers of an LVL beam are joined together with nails or bolts just as with a dimensional-lumber beam. You can purchase and handle lengths that reach the full distance between beam pockets, or at least halfway, minimizing butt joints.

LVL beams are dimensionally stable, so they can be set level with the mudsills without allowing for shrinkage. Also, LVL tends to be straighter and have no crown, so LVL beams don't take much tweaking to get them true.

On the downside, LVL beams are heavier than dimensional lumber, so plan your crew accordingly. A 36-ft. LVL beam that's 9½ in. high is more than two people can handle safely. LVL is also much more dense than regular lumber, which makes nailing the layers together more difficult. Pneumatic nailers don't always drive nails in completely. We find that clamping the members together before shooting the nails helps, and any nails that aren't sunk completely can be sent home with a hammer.

Long, straight, and strong comes at a price. Expect an LVL beam to cost twice as much as a dimensional-lumber beam.

Most center beams fit into recesses in the foundation called beam pockets. The simplest center beam spans from one side of the foundation to the other. With larger or more complex designs, there may be several beams, and some beams may span only a portion of the basement width.

Sketch the Beam before You Start

Before we order materials for a new house, we sketch a beam plan that shows the numbers and lengths of the boards in the beam. On the sketch we mark the centerpoint of each support column and the measurement between those points as well as to the inside edges of the foundation.

Most center beams are longer than the longest available stock lumber, so we plan for butt joints in each layer. With an engineer's approval, butt joints can fall between columns, but rather than take chances, we locate all butt joints over the columns.

These joints should be staggered between layers. On the sketch, we label each beam layer by number and the boards in each layer by letter to keep things organized on the job site.

Before we assemble the beam, we measure from the corners of the foundation and mark the exact center of the beam on the mudsills above the beam pockets. Working from the center, we draw the edges of each beam layer on the mudsill.

We also set up A-frame scaffolding to support the beam temporarily as we set it in place. The tops of the A-frames put the beam close to its final height, and staging planks on the lower cross bars put us in a good position for assembling and positioning the beam.

Keep the Beam Flat during Assembly

At this point, we cut the pieces of the beam to length. Each board is given its piece-and-layer

WEDGES KEEP THE BEAM STANDING UP. Temporary 2× wedges that are placed in the pocket keep the beam from moving while assembly continues.

CENTER BEAM GOES HERE. After the crew measures from the corner of the foundation, the exact location of the center beam and all its individual layers are marked on the mudsills.

UP, OVER, AND INTO THE POCKET. Preassembled beam sections are easy for two crew members to lift. Here, crew members slide one end of the beam into the foundation pocket and rest the other end on the A-frame scaffolding.

label, and the direction of the crown is marked. The boards are then spread out in their approximate location on the ground inside the foundation.

It's much easier to assemble part of the beam and lift it into position on the A-frames rather than build it from scratch in position. And although we've seen it done, we never preassemble the entire beam and try to lift it into place, which is a dangerous proposition regardless of the size of your crew. So starting at one end of the beam, we line up the second layer on top of the first. We keep the tops of the boards flush as they're nailed together.

Nails staggered every few feet are enough to hold the two layers together at this point. We try to keep the beam as flat as possible during this process. Any waves built into the beam as it's tacked together can be hard to take out later. For long center beams (more than 40 ft.), we assemble two or three of these two-layer sections, orienting the crowns in the same direction. The first of the assembled sections is then lifted onto the A-frames and slid into the beam pocket. Blocks of 2× wedge the beam section upright temporarily. Next we add preassembled sections or additional pieces until the first two layers are complete from pocket to pocket.

Brace the Beam to Keep It Straight

Before adding the rest of the layers, we brace the beam straight so that no curves are built in. We stretch a string the length of the partial beam, spacing the string from the beam with two short 1×3 blocks. A third block is used as a gauge.

Before nailing on the braces, we make sure the beam hasn't sagged. If it has, we adjust the A-frames

ALMOST A BEAM. The last board in the first layer completes the bridge between the foundation pockets. At this point, the first two layers are still only tacked together.

ONE PIECE AT A TIME. After the first two layers are braced straight and nailed off, each of the final layers is added one board at a time. After the entire layer is in place, the crew goes back and nails it off.

until the beam is approximately level again. Then we extend an adjustable 2×4 brace from the mudsills across the top of the beam at each support-column location. These handy braces, available through concrete-form supply houses, consist of a turnbuckle that is then attached to a 2×4. The adjustable ends of the braces are nailed 2 in. in from the edge of the mudsill so that they don't interfere with the rim joist and floor joists.

Keeping the beam roughly straight, we nail the other end of each brace to the beam. Then one crew member fine-tunes the brace on the turnbuckle end while another gauges the beam with a block. Besides keeping the beam straight, the braces keep it from rolling over while we add the final layers. When the two layers have been braced straight, we fasten them together permanently with rows of 12d or 16d nails every 12 in. to 16 in. The nails are driven at an angle, so they don't poke through the other side.

Add Final Layers One at a Time

The next layer can now be added to the beam with the crown up and the top flush with the rest of the beam. When that layer is tacked in place, we make sure the beam is still straight before nailing it off.

If the beam has a fourth layer, it is added the same way. As we add successive layers, we're careful to

keep the joints staggered and to install each board according to our sketch.

Set the Beam a Little High

Dimensional lumber always shrinks. A two-plate mudsill that's 3 in. thick can be expected to shrink ⅛ in. to ¼ in. over the first year or two as the house dries out. Even though we try to build our center beams of kiln-dried lumber, a beam can shrink up to ⅝ in. depending on its moisture content when it's installed. To compensate for this shrinkage, we

NEED A LIFT? Strings stretched across each support column keep the beam in plane with the sills. Screw jacks replace the scaffolding support, and using a gauge block, the crew sets the beam ¼ in. to ⅜ in. above the mudsills to compensate for future shrinkage.

install the beam ¼ in. to ⅜ in. above the top of the mudsills.

To level the beam, we first stretch a series of strings between the mudsills perpendicular to the beam, one string over each column location. We always use strong twine that can be pulled tight without sagging. Just as with straightening the beam, spacer blocks are placed under the strings at each end.

Screw jacks that raise the beam are then placed about 1 ft. from each column location, snugged up to the bottom of the beam. If screw jacks aren't available, there is a site-built alternative (see the sidebar below).

LIFTING A BEAM WITHOUT JACKS

HERE IS A SITE-BUILT ALTERNATIVE FOR raising a center beam without screw jacks. First, lay down an 8-ft. 2×6 in the basement on a flattened area of ground that runs perpendicular to the beam just to the side of the column location. Cut two 2×6s about 5 in. longer than the distance from the 2×6 on the ground to the bottom of the beam, and nail the two 2×6s together at the tip of one end with the other ends spread apart. Slide the nailed ends under the beam, and place the loose ends on the flat 2×6 to form an A-frame (see the photo at left).

Tack the top of the A to the underside of the beam, and tap the legs inward equally to make the beam rise. If more than one lift is being used, adjust each a little at a time and equally along the length of the beam until the beam is at the correct height. To keep the legs from kicking out, nail on 2× blocks to back up each leg.

INSTALLING THE SUPPORT COLUMNS

BEFORE THE COLUMNS CAN GO IN, WE FIRST make sure the beam is still level. If our strings have been stretched for more than a day, we retighten them to take out any sag. Then each post location is marked on the bottom of the beam, and the distance from the concrete footing to the bottom of the beam is measured at each post location.

If the layers on the bottom of the beam aren't flush, we chisel off a flat spot for the top column plate. We also take off any high spots on the footings for a smooth surface. The measurements are written on the beam. To obtain the actual cutting length, we subtract the thickness of the top and bottom plates if one or the other isn't already welded to the column, and mark that length on the column.

BEFORE YOU CUT

It's best to cut longer columns first so that if one is cut too short, it can be recut and substituted for one of the shorter columns. A large pipe cutter is the fastest and easiest way to cut a column, but it's an expensive tool to own and not always available to rent. An alternative is using a metal-cutting blade in a reciprocating saw or circular saw. But all cutting options begin with an accurate cutline around the circumference of the column.

The easiest way to draw the cutline is to wrap a large piece of paper around the column 1 , keeping an edge of the paper lined up on the mark and then matching the edge of the paper to itself as it wraps around the column. A pencil line follows the edge of the paper.

MAKE THE CUT

We gently cut through the metal skin of the column, following our line 2 . When we're most of the way through the steel, a light tap with a hammer breaks off the waste piece. If the concrete core of the column breaks off beyond the cutline, a few hammer taps on the concrete chips off the excess.

If the top plate hasn't been welded to the column, we fasten it to the underside of the beam. To let each column slip in more easily, we raise the beam about ⅛ in. with the screw jack. The column is then slid into place and set on the baseplate. Next we roughly plumb the column and lower the

jack until the column starts to bear the weight of the beam. The bottom of the column can then be tapped into place while it's plumbed with a level 3 . When the column is set, we make index marks around the baseplate just in case it gets bumped out of place during construction 4 .

SECURE THE COLUMN

If the slab hasn't been poured, the base of the column can be secured to the footing with a couple of masonry nails. If the slab has already been poured over the footings, we install lag shields into the concrete and then bolt down the baseplate of the column. As a final step, the column is welded to the top and bottom plates. We usually make a continuous weld around the ends of the column with a MIG welder. The small tabs on the plates should not be trusted to keep the column in place.

PAPER MAKES THE PERFECT CUTLINE. A sheet of paper with the edge set at the measurement and then wrapped around itself creates a continuous line for cutting around the column.

THAT'S A BIG PIPE CUTTER. A large pipe cutter is the easiest and fastest way to cut a steel column to length. A circular saw or reciprocating saw with a metal blade can also be used.

PERFECTLY PLUMB. After the support column and plates are inserted under the beam, one crew member taps the column into plumb while another keeps his eye on the level.

COLUMN BASE BELONGS HERE. After the column has been plumbed, the position of the column base is marked on the footing in case the column is bumped before the basement slab locks it into place.

BEAM ENDS REST ON PERMANENT SHIMS. Shims that support the ends of the beam can't rot or be crushed. Steel plates and plywood can be used for small spaces (far left), and lumber shims with the grain in a vertical position are used to fill larger spaces (left).

With the screw jacks in place, the A-frames can be removed. We adjust the jacks until the beam is ¼ in. to ⅜ in. higher than the mudsills, nudging each jack a little at a time to bring the beam slowly up to position.

Permanent Shims in Beam Pockets

Once the beam is at the correct height at each of the column locations, we move to the beam pockets and install permanent shims to lift the ends of the beam to the same height. Shims should be made of dimensionally stable material that won't crush or rot.

The thickness of the space below the beam affects the choice of shim material. Spaces ½ in. or less are best filled with steel plates. Several plates can be stacked to fill the void, or steel plates can be used in combination with plywood (see the photo above). Softwood shims, such as cedar shingles, should never be used to shim a center beam.

Pressure-treated lumber shims should never be used with the grain flat. In this position, they can shrink and allow the beam to settle. Instead, we install these shims with the grain in a vertical position. These shims work best to fill spaces 3 in. or more.

We also replace the temporary blocks on the sides of the beam pockets with pressure-treated blocks. To make sure that the beam stays on its layout marks while the floor is being framed, we nail steel straps in an X between the beam and the mudsill. We usually make the X out of the steel strapping from the lumber delivery.

RECYCLED STEEL STRAPPING KEEPS THE BEAM IN PLACE. An X made of steel strapping from the lumber delivery anchors the beam to the mudsills while the floor is being assembled.

Columns: Now or Later?

Most of the houses we build call for 3½-in.-dia. concrete-filled steel columns (also known as Lally columns) as permanent supports. The columns can be installed at this point, but because there is no weight on the beam yet, we sometimes wait until the floor system has been installed. When they're supporting the weight of the floor, they're less likely to be knocked out of place accidentally.

But don't wait too long. Once load-bearing walls and the weight of a second floor have been added, it may not be possible to lift the beam without more powerful jacks.

Frame a Strong, Stable Floor with I-Joists

BY JOHN SPIER

Manufactured I-joists are used in about 45 percent of new wood-frame construction, and that amount is expected to rise. I can't remember the last time I framed a new floor with dimensional lumber. To me, I-joists make more sense. They are straighter, stronger, and lighter, and they span longer distances than ordinary 2×s. They are also a more-efficient use of resources because they can be manufactured using lesser-quality trees. Of course, I-joists cost a bit more, but they also are much easier to install, meaning I get a big savings in labor.

Then again, nothing is perfect, and I-joists have a few disadvantages compared to standard dimensional lumber. They don't cope well with careless handling, they are more sensitive to moisture, and they shouldn't carry any load until they are fully sheathed. They also aren't as amenable to job-site change orders, and many lumberyards don't stock them.

I-Joists Are Part of a Carefully Planned Floor System

With I-joists, system is the operative word. A floor framed with I-joists is designed as a package, with all its components specifically located in the overall structure. This is different from conventional fram-

ing, where joist size is selected based on maximum span and then is used for an entire floor. In an I-joist floor, components have the same depth, but flange widths, joist spacing, and attachment details can vary throughout the system to make the most efficient use of materials.

ALIGN JOISTS AGAINST A SNAPPED LINE

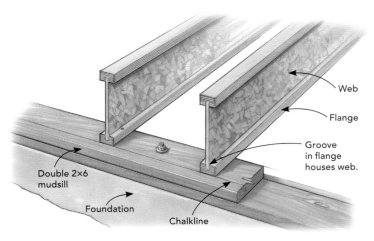

Web

Flange

Groove in flange houses web.

Double 2×6 mudsill

Foundation

Chalkline

The 1¼-in.-thick engineered rim joist can swell.

SNAP LINES. In most cases, I-joists are nailed to the mudsill before the rim joist is installed. Assuming everything is squared up in advance, I start by snapping a chalkline 1⅜ in. from the outside edge of the mudsill. This allows enough room for the 1¼-in.-thick engineered rim joist to swell, which it always seems to do.

MEASURE THE DISTANCE from the beam (or other terminating point) to the chalkline on the mudsill at each joist location, subtract about ⅛ in., and mark each length on the mudsill. The ⅛-in. gap will be on the hanger end of the joist and helps to prevent squeaks in the finished floor.

The first step in building an I-joist floor is having it designed. I provide lumber suppliers with complete sets of building plans, which they pass along to one or more engineered-lumber vendors. In a week or two, each vendor sends me floor-framing plans and quotes for the I-joists, beams, blocking, hardware, and other material.

It's important to review the engineer's floor-framing plans carefully, especially to make sure that they work with the builder's plans. Sometimes the engineers miss a key element and design a framing plan that doesn't accommodate the plumbing or the ductwork, or they might specify some details that experience has taught me to avoid. On the other hand, an engineer sometimes simplifies and improves the structure by coming up with a framing plan that I hadn't envisioned.

I also review the attached materials list carefully. If the engineer has specified a lot of mixed, short-length pieces, I often combine them into longer lengths that can be cut to length on site. The longer lengths are easier to handle, reduce waste, and give me a margin for error if I make a cutting mistake later or find a damaged piece in the pile. Blocking panels, which are short lengths of joist used to transfer loads over bearing walls, are often listed as a pile of separate 2-ft.-long I-joists, but I can save

GANG-CUTTING I-JOISTS WORKS BEST

BUNDLES OF I-JOISTS OFTEN arrive on site with one end factory square and the other end looking as if it were cut by an unemployed logger with a dull chainsaw. Because I usually need to cut the joists to length anyway, I keep the factory-square ends together and cut off the other ends.

I find that I-joists are ideally suited for gang-cutting. As shown at left, I can set them up on edge on a pair of sawhorses, align the ends, square a line, and use a circular saw to cut through the flanges along one edge of the stack. Then I roll each joist down and use the kerf on the flange as a reference to finish the rest of the cut. I cut 9½-in. and 11⅞-in. I-joists using a large Speed Square® as a saw guide. Wider joists need to be marked with a framing square. If the flanges are too wide for the cutting depth of my circular saw, I roll the saw over the edge of the I-joist to finish the cut (see the photo at right).

myself some money by getting these short pieces out of cutoffs instead.

I also always buy one extra of the longest joist just in case; if I don't use it, I can return it. I check the beams, too, because I often can combine them into continuous lengths that are stronger and can be installed faster. Finally, I check the hardware list. If the engineer has specified hardware items that I'm unfamiliar with, I look them up in the catalog to see how they are installed.

Snap Lines, Install Beams, Then Move to the Floor Joists

All floor framing starts with, and depends on, mud-sills or wall plates that are straight, level, and square, so check for these conditions first. I start an I-joist floor by snapping layout lines along the mudsills and across any intermediate walls or beams. This step is easier to do when nothing is in the way, and it helps to keep the work straight later. Remember to recheck the plans for any interference with toilet flanges or tub and shower drains. I-joists can't be easily notched or headed off later.

If big beams are incorporated in the floor system, I install them next because they are easier to move into position and brace securely before the joists get in the way. Often, beams and headers are used to form openings such as stair or chimney holes, which come next. Once I install these components, I can get to work on the I-joists.

Because these joists are fairly stable when nailed to the plates and sills through their flanges, I can

ATTACH THE HANGERS TO THE JOISTS

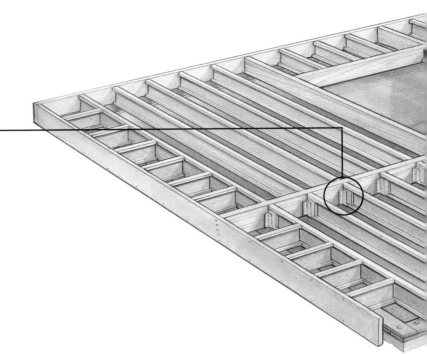

INSTALL HARDWARE. For I-joists that require metal hangers on one end, I install the hardware on the joist right after I cut the joist to length. A metal-connector nailer like the one shown here (www .bostitch.com) really speeds up this process. For this job, I had to install web stiffeners on each I-joist before attaching the hanger. I made the stiffeners from plywood offcuts.

ANGLE NAILS INTO SILLS AND PLATES

FASTEN I-JOISTS. I-joists that are not carried in hangers are fastened to the mudsill or top plate with nails driven at an angle through the flanges. If the joists will be hung from a beam on one end, one person supports the hardware end of the joist and tacks it in place with a couple of nails. The person near the mudsill confirms that the end of the I-joist lands on the chalkline before both ends are permanently fastened.

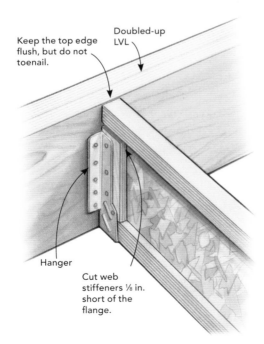

Keep the top edge flush, but do not toenail.

Doubled-up LVL

Hanger

Cut web stiffeners ⅛ in. short of the flange.

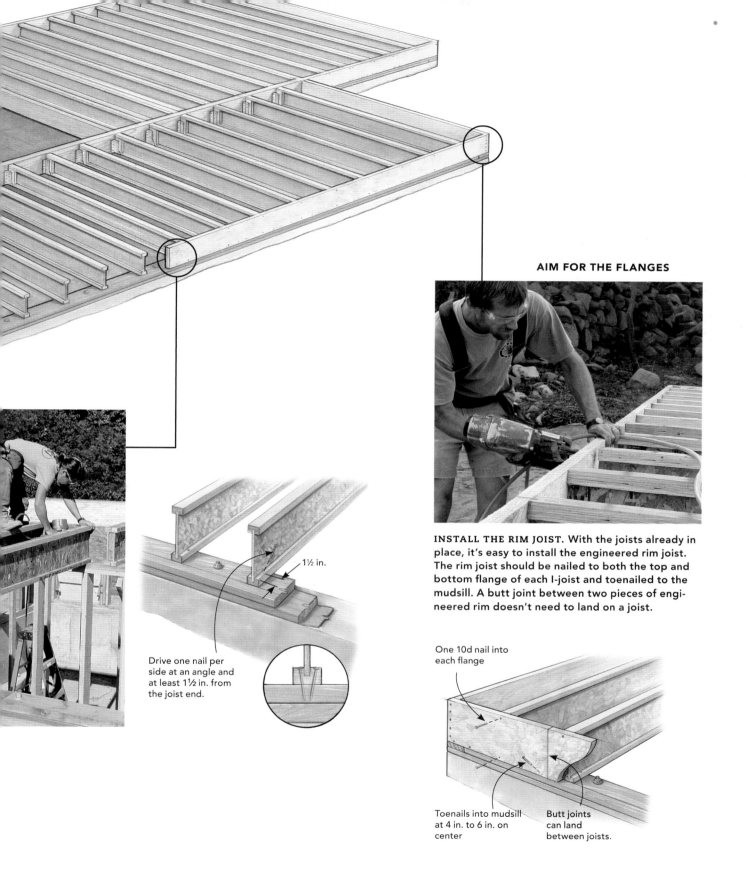

AIM FOR THE FLANGES

1½ in.

Drive one nail per side at an angle and at least 1½ in. from the joist end.

INSTALL THE RIM JOIST. With the joists already in place, it's easy to install the engineered rim joist. The rim joist should be nailed to both the top and bottom flange of each I-joist and toenailed to the mudsill. A butt joint between two pieces of engineered rim doesn't need to land on a joist.

One 10d nail into each flange

Toenails into mudsill at 4 in. to 6 in. on center

Butt joints can land between joists.

REINFORCE WITH BLOCKING PANELS

Blocking panels are short lengths of I-joist used to transfer loads and to stiffen the floor assembly.

BLOCKING PANELS ARE INSTALLED between an I-joist and a rim joist that is cantilevered beyond the foundation. I cut the blocking panels to fit and secure them by driving toenails into the flanges. There is no need to notch the blocking panels to fit against the webs, just the flanges. The drawing at right shows a more typical blocking panel installation at a cantilevered bay window. The engineering for any type of cantilever will be done for you.

Do not nail the top flange; let the subfloor tie it together.

install them before the rim joists. If I'm working on mudsills, I snap a line on one sill and hold the joist ends to it. If I'm building on a wall that has been braced straight, I scribe a line on the top plate.

Joists that attach to a flush beam or a header using hanger hardware can and should be cut a bit short; ⅛ in. is about right. The hangers are forgiving, and tightly fitting joists can push things out of alignment. They also tend to become squeaky spots in the finished floor.

I-joists that require web stiffeners and hangers on one or both ends can be cut and prepped assembly-line style (see the sidebar on p. 119). I set up a couple of heavy-duty sawhorses with air hoses and nail guns

at each end. This way, I can gang-cut the joists, install the necessary stiffeners or hardware, and move the joists into position. I generally cut and prep the joists in sets, then install them in the same order.

The Rim Joist Goes on after the Joists Are in Place

Unlike I-joists, engineered LSL (laminated strand lumber) or LVL (laminated veneer lumber) rim joists are not delivered in specific lengths. They simply show up as a bunch of stock pieces.

Engineered rim-joist stock is usually 1¼ in. thick. But it often swells, so I allow 1⅜ in. for it. Besides, it's much easier to tap a rim joist out a bit than it is

A SUBFLOOR COMPLETES THE SYSTEM

ATTACH SQUASH BLOCKS WHERE NEEDED

TO KEEP I-JOISTS FROM BEING OVERLOADED by concentrated loads from above, the engineer's floor-framing plans require dimensional-lumber squash blocks in key areas of the floor.

COMPLETE THE SYSTEM WITH THE SUBFLOOR. Every floor plan is different, but for this project it made sense to snap a chalkline 46 in. from the outside edge of the rim joist and start the first row there. After I lay a thick bead of construction adhesive on the top edge of each joist, the subfloor panels are laid out and tacked in place; butt joints should land over the top of a joist. Once the first row of subflooring is tacked in place, I snap a chalkline along the outside edge and rip off the overhanging tongue so that the panels are flush to the rim joist. I go back and finish nailing or screwing each panel every 6 in. on center. A sledgehammer comes in handy for persuading sheets to fit together tightly, but I hammer against a scrap block to protect panel edges from damage. As shown above, a little boot pressure is helpful to align the joists with the layout marks before fastening.

Unlike installing a subfloor over a dimensional-lumber frame, an engineered system requires that the edge of the tongue-and-groove subfloor panel be full thickness, flush with the outside edge and in full contact with the rim joist below.

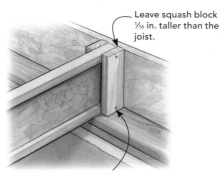

Leave squash block ¹⁄₁₆ in. taller than the joist.

Drive one 10d nail into each flange.

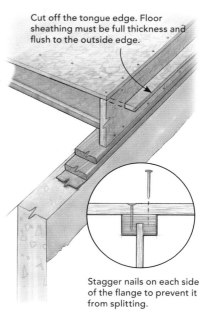

Cut off the tongue edge. Floor sheathing must be full thickness and flush to the outside edge.

Stagger nails on each side of the flange to prevent it from splitting.

LOCATE HOLES AND NOTCHES WITH CARE

IN CONVENTIONAL FRAMING, THE SIZE and location of holes cut through the floor joists for running utilities are defined by the building code. I-joists are different. Except for cutting to length, the flanges of an I-joist can't be cut, drilled, or notched. Most I-joists come with knockouts punched in the webs for running pipes and wiring. If you need to cut additional holes, they must be located in the web of the I-joist and must meet the manufacturer's guidelines, which vary from project to project. This information will be provided to you, typically in the form of a pamphlet attached to the I-joists.

Knockout

to trim the end of an I-joist to eliminate a bump in the building. Rim joists need to be nailed according to a fastening schedule, typically with one 10d nail into each joist flange and toenails into the mudsill or plate every 4 in. to 6 in. I use full-length rim-joist pieces in the corners and fill in the middle with shorter lengths.

Joints between pieces of engineered rim joist can be butted together anywhere, even between I-joists. A word of caution: Nailing through the rim joist and into top flanges can be dangerous. The target is small, and there is no good alternative to having one hand in the line of fire. Awareness, caution, and a sequentially firing nail gun are helpful here. After the rim joists are in place, install the blocking, squash blocks, and any necessary cantilever reinforcements (see p. 122).

A Subfloor Locks the Joist Layout and Finishes the Floor

Installing a subfloor over an I-joist frame is almost the same as installing it over a conventional frame, but there are a few important differences. Unlike a conventional floor, where the starting edge of the plywood subfloor is held back ¼ in., the subfloor over an engineered rim needs to be full thickness (no tongue) and flush to the outside edge.

Because I have to cut off the tongue and start with a ripped sheet anyway, I check the overall width of the floor and the locations of large jogs or openings. I often can plan a starter-course width that allows for efficient use of rips and cutoffs.

Engineered subfloor panels usually have layout marks that you can use, but be careful not to miss any of the small shifts in joist spacing that you might have made for plumbing or other details. I run most of my subflooring long over the ends and into the openings, snapping lines and trimming it later. Also, be aware of any top-flanged metal joist hangers when you're cutting subflooring in place. I leave many of the edges and ends unnailed until they are trimmed straight. This enables me to make delicate adjustments with a sledgehammer where needed.

Supporting a Cantilevered Bay

BY MIKE GUERTIN

When a client wants to add curb appeal to a new home, I dip into my Mr. Potato Head® bag of tricks: A distinctive window here, a reverse gable there, fancy trim details, an entry portico or a porch—and voilà! It's enough to make an architect cringe.

One of the best-selling upgrades is an angled bump-out or bay. It adds a few square feet, creates a distinctive room inside, and dresses up the home's exterior. Although I could just install a bay window for light and effect, I find the floor-to-ceiling bay more appealing as well as competitive in cost.

But a bay is only as strong as the floor that it's built on. Here I'm going to concentrate on the proper techniques for framing the cantilevered floor that supports a bay. For this project, the bay was 8 ft. wide and extended 2 ft. from the house. The sides of the bay were set at 45 degrees, but they could have been set at any angle.

PLATES BEFORE JOISTS. Before the cantilevered-bay joists are cut and installed, the wall plates are cut and laid out. At this point, all measurements are checked, and the exact width of the bay is established.

Cantilevered Joists Save Foundation Work

Cantilevering the bay costs less than an angled foundation. It's also easiest to frame one of these bays when the joists run parallel to the floor framing. In this scenario, the common joists are just lengthened to form the bay joists, eliminating the need for headers and hangers. But I wasn't so lucky on this project. The floor joists of this bay ran perpendicular to the main joists (see the photo below).

The cantilever wouldn't be carrying any loads but the bay itself, so I followed the two-thirds in, one-third out cantilever rule of thumb. With a 2-ft. cantilever, the bay joists would be anchored to a tripled floor joist 4 ft. in from the outside of the house. But I waited to add the second and third joists until just before sheathing the deck. Having only one common joist allowed me to nail through it to attach the bay joists initially. The bay joists follow the 16-in.-on center layout regardless of exactly where the bay is placed, so first I put in all the 4-ft. joists that fell on each side of the bay area.

Cut and Lay Out the Bay Plates First

Before I laid out the exact location of the bay, I cut and laid out the top and bottom plates for the bay walls. Although this step may seem a bit premature, I always want to be certain that the windows will fit and that I'll still have room inside and out for the trim. The plates also help me to figure the length and cut for each joist.

A little basic math and a calculator gave me the plate lengths. With the bay cantilevering 2 ft. and the walls at 45 degrees, I needed to come in 2 ft. from each side for the bay's front plate. With 22½-degree angles on each end (half of 45 degrees), I cut the plate for the bay's front wall at 4 ft. from long point to long point.

BAY JOISTS HANG OFF THE MAIN JOIST When the bay joists run perpendicular, they are nailed to a main joist. After the cantilevered joists are attached, the main joist is tripled and joist hangers are installed.

With some help from Pythagoras, I cut the side plates, again with 22½-degree angles on each end and with the outside face measuring 33¹⁵⁄₁₆ in. long point to short point (short point because the adjoining wall plate is also cut at 22½ degrees to form the inside corner). With the plates laid out on a flat surface, I marked the rough opening for the window centered on the 4-ft. plate.

To get the width of the trim (exterior and interior) to match on both sides of the bay's outside corners, I make sure that the distance is the same from the corners to the edge of the rough openings for all three windows. After the rough openings are marked out, I also make sure that I have enough space left (at least 1 in.) for the inside-corner trim.

Center the Bay on the Interior

I usually center a bay on the room inside. In this case, that threw it slightly off-center on the exterior, but the difference wouldn't be noticeable. I marked the location of the 8-ft. opening on top of and on the outside face of the sill plate. (On this house, the sill plate is actually the top plate of a framed wall for a walk-out basement.)

Next, I marked the outside corners of the bay on the sill, showing me which joists would cantilever the full distance. The house's rim joists were then run to the locations of the first cantilevered joists inside the 8-ft. layout marks rather than being mitered into the bay's rim joists. The extended rim joists are nailed square to the cantilevered joists to hold them

SETTING THE OVERHANG. With a 2-ft. cantilever, the longest joists overhang 22½ in. The joists are then tacked in place.

BAY STARTS HERE. After the bay joists are tacked in place, the outside edge of the bay is carried up from the face of the sill onto the rim joist.

straight and plumb, and in turn they provide a more solid place to secure and plumb the bay's rim joists where they meet the house wall.

The cantilevered bay joists were put in next. The joists that fell in the middle 4-ft. section of the bay cantilevered by 22½ in. (2 ft. less the thickness of the rim; see the top photo on p. 127). These joists were tacked to the sill and nailed to the common joist. The joists that fell on the angled sidewalls were left a little long and cut to exact length later.

When all the joists were nailed in, I ran a string to straighten the main rim joist and then drew square lines up from the lines I'd made earlier on the sills, indicating the outside edges of the bay. The rim joist for the outer wall of the bay was then cut and nailed in, left long to be cut to exact length later.

OUTER RIM JOIST IS LEFT LONG. Rim-joist stock is nailed to the cantilevered joists and left long until cutlines are transferred from the plates.

PRECUT WALL PLATES PROVIDE THE SHAPE OF THE BAY. The bay plates are laid on top of floor joists and rim joists that were run long and needed to be cut.

Use Plates to Mark the Joist Cuts

With the joists in place, I next set the bay's wall plates in position over the cantilevered joists and rim. At the outside corners of the plates, I squared down a cutline indicating where to miter the bay's rim joist (see the photo at right).

A line was also drawn along the outside edge of the sidewall plates onto the tops of the joists that were left long. This line is the perimeter of the bay, so holding a 2× block inside the line and drawing a second line allowed for the rim joist and gave me the actual cutline. After squaring down the cutlines, the joists were trimmed at 45-degree bevels. If the amount of waste is more than about a foot, I rough-cut the length so that the cutoff isn't heavy and unwieldy. Finally, I cut, fit, and installed the angled rim joists.

CUTLINES ARE SQUARED DOWN FROM THE PLATES. Lines are extended down from the corners of the plates for the rim-joist cuts, and the lengths of cantilevered side joists are marked 1½ in. from the outside edge of the plate.

CUT DOWN WHERE THEY STAND. The cantilevered side joists and the outside rim joist are cut to length in place.

INTERLOCKING JOIST CONNECTION. The main rim joist on the house is run to the first cantilevered joist and nailed, keeping both joists plumb and square and providing a solid landing place for the angled rim joist.

WAIT—THOSE JOIST HANGERS ARE UPSIDE DOWN! To prevent uplift, joist hangers are nailed on top of the bay joists as well as on the bottom, where they support the weight of the floor.

With all the floor framing for the bay complete, I tripled the main joist that the bay hung from and installed two joist hangers on each of the cantilevered joists, one right side up and the other upside down. The theory is that the upside-down hanger prevents the joist from lifting from the weight of the cantilever. Some framing crews block between the cantilevered joists at the sill plate, but I prefer to leave the space open to slide in insulation later.

With floor framing done, I finished sheathing the main deck, extending the sheathing over the cantilever. I left the walls and roof of the bay until the entire house was framed. The house walls give me a nice surface for attaching the bay roof, and a roomy exterior deck wrapped around the bay, disguising the fact that it was sitting on sturdy cantilevered floor framing.

SHEATHING COMPLETES THE FLOOR FRAMING. With the bay joists on regular centers, the deck sheathing can be cut and extended to include the bay without changing the sheathing pattern.

6 Ways to Stiffen a Bouncy Floor

BY MIKE GUERTIN AND
DAVID GRANDPRÉ

If you haven't fallen into the basement of your house already, don't worry; your bouncy floor is probably not an indication of a disaster waiting to happen. But it is a good idea to understand why your floor is bouncy before planning your upgrade. You want to find out if the condition is due to wood decay, insect damage, or any other external force at work that is causing the floor structure strength to be reduced.

Floor deflection is common in older homes because the floor joists often are smaller or are spaced farther apart than the joists in modern homes. Of course, new homes also can have bouncy floors if the joists are approaching the maximum spanning distance for the weight they are supporting. Long-span joists may meet design criteria and the building code, yet still feel uncomfortable.

A well-designed wood floor feels stiff as you walk on it but still gives slightly under foot, absorbing some of the impact of your steps. Too much bounce, though, can make the china cabinet wobble. You can shore up floor joists and reduce the bounce in a number of ways, but the six methods outlined here represent a mix of common and not-so-common solutions. The best choice depends on access to the joists, obstructions in the floor system, or current remodeling plans; one technique or a combination may be your most practical solution.

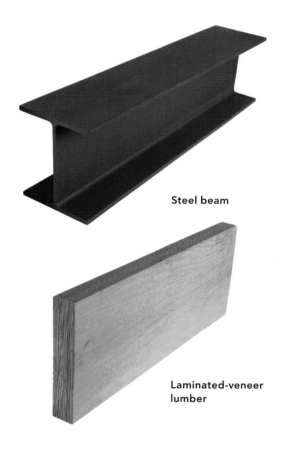

Steel beam

Laminated-veneer lumber

STEEL OR LVL BEAMS can be used in lieu of dimensional lumber. Their added strength will allow for wider column spacing, but larger footings may be required to carry the load.

131

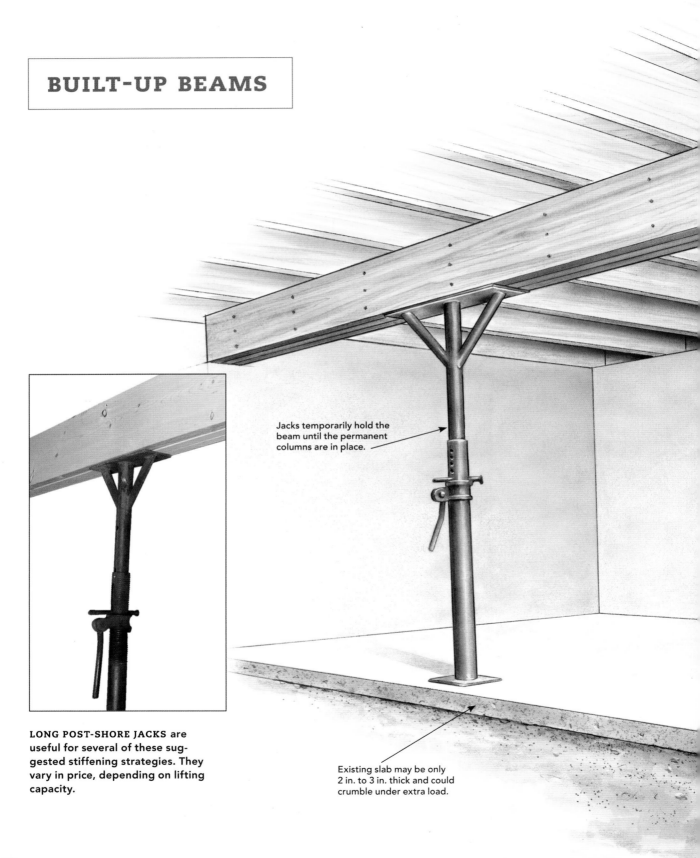

BUILT-UP BEAMS

Jacks temporarily hold the beam until the permanent columns are in place.

LONG POST-SHORE JACKS are useful for several of these suggested stiffening strategies. They vary in price, depending on lifting capacity.

Existing slab may be only 2 in. to 3 in. thick and could crumble under extra load.

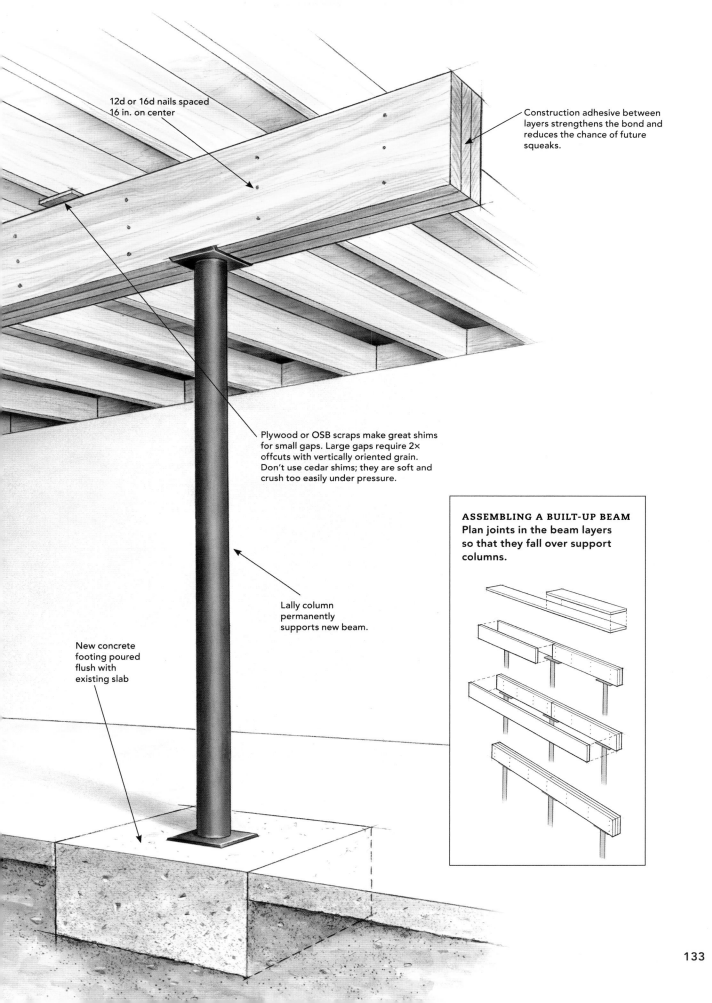

12d or 16d nails spaced 16 in. on center

Construction adhesive between layers strengthens the bond and reduces the chance of future squeaks.

Plywood or OSB scraps make great shims for small gaps. Large gaps require 2× offcuts with vertically oriented grain. Don't use cedar shims; they are soft and crush too easily under pressure.

Lally column permanently supports new beam.

New concrete footing poured flush with existing slab

ASSEMBLING A BUILT-UP BEAM
Plan joints in the beam layers so that they fall over support columns.

It's important to make these improvements carefully. If existing joists have been weakened due to rot or insect damage, glue and fasteners won't hold well, and your work may be ineffective. Loose blocking or an underfastened subfloor won't bring any benefit, so take extra time and care during installation. Also, you can use jacks to relieve the load on joists while the work is being done. Jacks improve the effectiveness of your floor-stiffening work.

1. Built-Up Beams Are Rock Solid but Reduce Headroom

This solution works best in crawlspaces where you aren't too concerned about limiting headroom or cluttering the space with columns. If you don't mind the obstructions or loss of headroom, though, beams and columns certainly can be added in basements, too.

The important thing to remember when adding a support beam is also to add proper footings to support each column. In most instances, a 2-ft.-square, 1-ft.-deep footing provides adequate support. However, when you're installing LVL or steel beams with wider column spacing, larger footings may be necessary to support the load.

The beam size depends on the load and span of the beam between columns. Also, keep in mind that the closer you space the footings and columns, the more rigid the new beam will be and the stiffer the floor will feel.

To make a new footing, cut the slab, dig out the earth beneath, and pour concrete flush with the top of the slab. Next, snap a chalkline across the underside of the joists in the middle of the span to help align the new beam. Use post shore jacks, screw

SISTER JOISTS

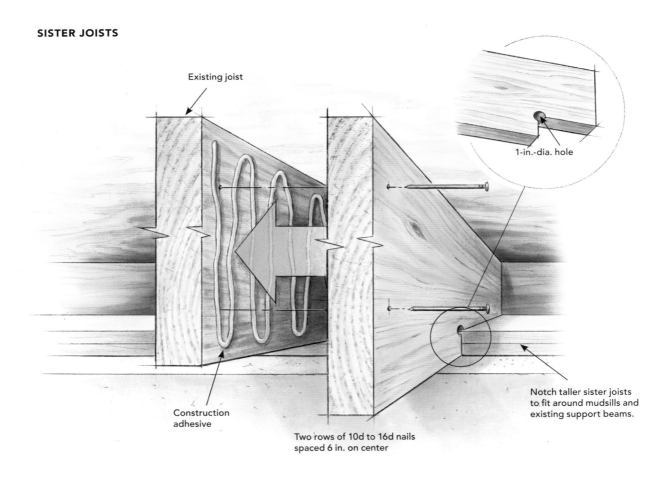

Existing joist

1-in.-dia. hole

Construction adhesive

Two rows of 10d to 16d nails spaced 6 in. on center

Notch taller sister joists to fit around mudsills and existing support beams.

jacks, or hydraulic jacks to lift the new beam into position beneath the joists. Finally, cut and install new columns to fit between the beam and the new footing.

2. Stiffening the Floor with Sister Joists Is a Tried-and-True Method

Adding a second joist of the same size alongside each existing joist, also known as sistering the joists, stiffens a floor. When headroom permits, sistering with taller joists provides more bang for the buck than sistering with same-size joists. Even though taller joists need to be notched to fit existing mudsills and support beams, the added depth along the middle of the span provides extra support and further reduces bounce.

Engineered lumber—LVLs, for example—also can be used as sister joists and adds more stiffness to a floor than dimensional lumber.

If the existing floor joists are bowed downward noticeably, they might need to be jacked up slightly to make installing the new joists easier.

To minimize future squeaks, spread construction adhesive onto both the existing joist and the new joist. Position the top of the new joist alongside the top of the existing joist. Use a sledgehammer or a pry bar to force the bottom of the new joist along the mudsill and center support beam of the floor system (or the opposite mudsill on short spans) until it's flat against the existing joist. Nail the new joist to the existing joist with two rows of 10d to 16d nails spaced 6 in. on center. For additional stiffness, sister joists can be applied to both sides of a joist.

OVERCUT NOTCHES ARE PRONE TO SPLITTING. Avoid runout by drilling a 1-in.-dia. pilot hole and then cutting up to it.

MAKE IT FLAT. Use a sledgehammer to force the bottom of the new joist along the mudsill and center support beam of the floor system.

3. Flexible Plywood Strips Are Great for Tight Quarters

An alternative to adding full-length sister joists is to apply two layers of plywood to one side of the bouncy joists. Shorter, lighter, more flexible plywood strips often are easier to install in tight quarters than full-size dimensional lumber.

Rip ¾-in. plywood into 8-ft.-long strips equal to the height of the existing joists. Use a combination of construction adhesive and nails or screws to fasten two layers of strips to the existing joist, spanning from the mudsill to the center support beam or opposite mudsill. Place the first 8-ft.-long strip 1 ft. off center from midspan, and fill toward the ends with pieces. Then glue and fasten a second layer of plywood starting with a full strip 2 ft. off center from the first layer.

When fastening to joists less than 8 in. tall, drive two rows of 8d nails or 2-in. wood screws 6 in. on center into each layer of plywood. If the joists are taller than 8 in., add a third row of fasteners.

This system relies on good workmanship for success. Be generous with adhesive and fasteners, and make sure that the existing joists are solid and not deteriorated. If the connection between the two plywood layers and the existing joists isn't solid, you don't maximize the benefit of using this technique.

PLYWOOD STRIPS

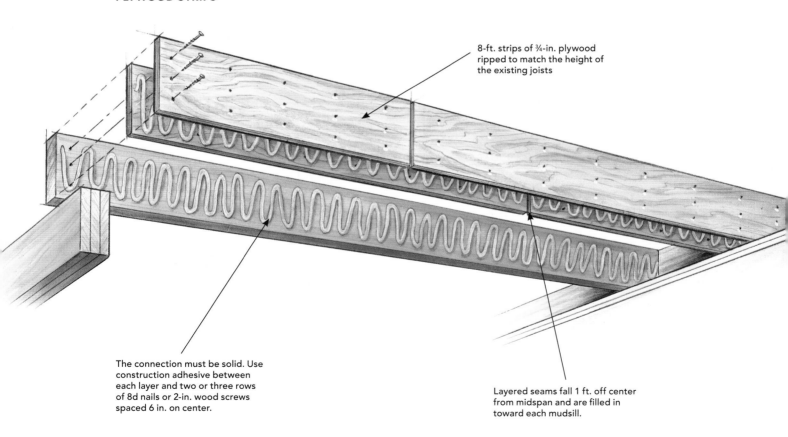

8-ft. strips of ¾-in. plywood ripped to match the height of the existing joists

The connection must be solid. Use construction adhesive between each layer and two or three rows of 8d nails or 2-in. wood screws spaced 6 in. on center.

Layered seams fall 1 ft. off center from midspan and are filled in toward each mudsill.

4. Fix Weakened Planks and Second-Floor Bounces with a New Layer of Plywood

Floors in older homes often are decked with diagonally laid 1× planks instead of plywood or OSB. If you're remodeling or just adding new flooring to a room, consider adding a layer of ¾-in. plywood subfloor sheathing over the lumber decking. When nailed through the old subfloor and into the joists, the new subfloor can help to reduce floor bounce. This solution also works for problematic second floors, where accessing the joists through the first-floor ceiling isn't a possibility.

When considering this option, think about the transition between old floor heights and new floor heights at doorways. You may need to add thresholds and cut doors. Also consider the loss of headroom.

Sheathing also can be applied to the bottom of the floor joists instead of or in addition to being installed on top.

To minimize the chance of future squeaks and to help secure the new floor sheathing, start by spreading a generous amount of construction adhesive over the old planks. Lay the new sheets perpendicular to the floor joists, and orient panel ends over joists. Choose ring-shank nails or screws that are long enough to penetrate the joists by 1½ in., and space them every 6 in. on center. Add an additional row of fasteners midway between the joists to pull the layers tightly together.

When installing on the underside of the joists, the process is the same. You can apply sheathing to the underside of floor joists only if the bottom edges are at the same level. If the joist level varies by more than ½ in., you can't use this method.

NEW LAYER OF PLYWOOD

Apply construction adhesive to the subfloor before additional plywood is laid.

Added subfloor thickness may require transition strips or trimming of doors.

Nails or screws spaced 6 in. on center should penetrate 1½ in. into the existing joists.

Panels should be perpendicular to joists with seams landing on joist centers.

A row of fasteners between each joist tightens the layers and reduces the chance of future squeaks.

**Perforated
steel strap**

5. Metal Straps Are Deceptively Effective and Simple to Install

Tim Brigham of Koloa, Kauai, Hawaii, suggested the use of continuous steel straps to stiffen joists. Wrap the joists from the top of one end, around the bottom at midpoint, and back to the top of the opposite end. When a load is applied to the middle third of the joists, the steel strap transfers the force of the weight to the nails along the length of the joist, particularly along the ends where the joist is more rigid.

To begin, lift up the joists with jacks and a temporary beam about 1 ft. from midspan of the joists. Raising the joists slightly (anywhere from ⅛ in. to ½ in., depending on the span and on the conditions) before installing this system helps to ensure that the straps are nice and tight when the jacks are removed.

Starting above the mudsill on one side of the joist, sink metal-connector nails through every hole along the first 2 ft. of strap. With the first 2 ft. fastened securely, the strap then should be fastened every 6 in. and folded with crisp bends around the bottom of the joist and onto the opposite side. Nail the strap to the opposite side of the joist the same way as the first side. For additional support, straps can be installed on both sides of the joists and cross-lapped at the center. Once all the joists are strapped, remove the temporary beam.

METAL STRAPS

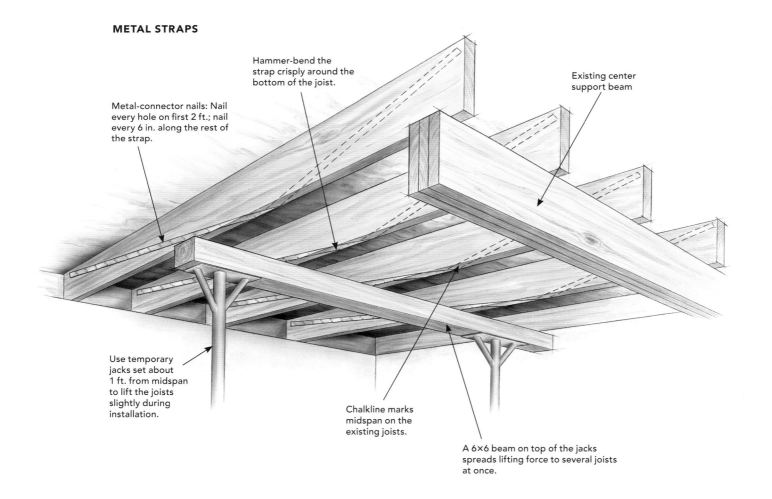

Hammer-bend the strap crisply around the bottom of the joist.

Existing center support beam

Metal-connector nails: Nail every hole on first 2 ft.; nail every 6 in. along the rest of the strap.

Use temporary jacks set about 1 ft. from midspan to lift the joists slightly during installation.

Chalkline marks midspan on the existing joists.

A 6×6 beam on top of the jacks spreads lifting force to several joists at once.

6. Solid Blocking Ties the Floor Together but Takes Time to Install Correctly

Properly installed solid-wood blocking helps to transfer weight to adjacent joists so that the floor acts as a stronger unified system. If you already have blocking or bridging installed between joists, it may be ineffective because it's not tight. Metal and wood cross-bridging are both prone to loosening over time as wood expands and contracts. Solid blocking is susceptible to shrinkage, but it typically works better than using the cross-bridging method.

This method works best if joists are dry, so it's a good idea to wait until late winter or early spring when the heating season is coming to an end and the moisture content of the joists likely will be at its lowest. Start by snapping a chalkline at the middle span of the floor running perpendicular to the joists. Using dry dimensional lumber, cut blocks just a whisker longer than the space between joists, and pound them into place so that they are tight. Blocks that aren't tight will end up causing squeaks.

SOLID BLOCKING

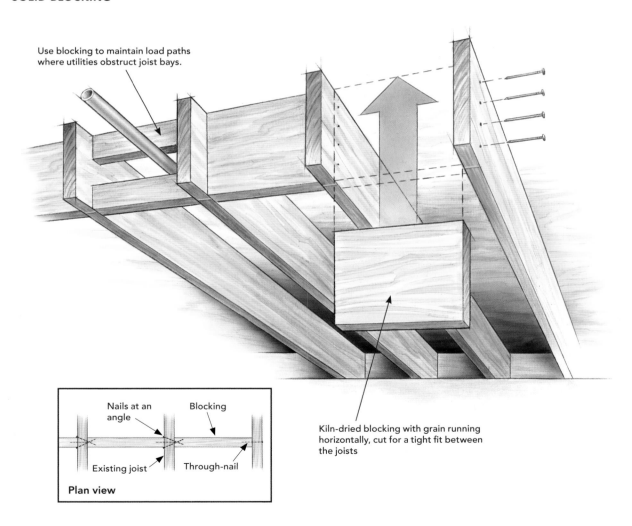

Use blocking to maintain load paths where utilities obstruct joist bays.

Kiln-dried blocking with grain running horizontally, cut for a tight fit between the joists

Nails at an angle

Blocking

Existing joist

Through-nail

Plan view

SOURCES

ABLE BUILDERS EQUIPMENT, LLC
Long post-shore jacks
www.ablebuilders.com

**SIMPSON STRONG-TIE
COMPANY INC.**
Simpson Strong-Tie CS20
Perforated Steel Strap
www.strongtie.com

It may be tempting to install this type of blocking in a staggered line because it's easier to fasten. But solid blocking is meant to work as a system, so keep the blocks in line. If you need to get around pipes or ductwork, use split blocks (a pair of 2×4s, for instance) on top and bottom to maintain the path.

Through-nail into the end of each block using three or four 16d spikes or 3½-in.-long wood screws. Pneumatic palm nailers work great for driving nails in these tight situations. For added support, you also can install two or three rows of blocking spaced equally apart.

Pneumatic palm nailer

A PNEUMATIC PALM NAILER is great for driving nails in spaces where swinging a hammer or fitting a full-size framing nailer isn't convenient.

PART 4

Framing Walls

Careful Layout for Perfect Walls

BY JOHN SPIER

Framing walls is one of the most fun parts of building a house. It's fast, safe, and easy, and at the end of the day it's satisfying to admire the progress you've made. Before cranking up your compressor and nail guns, though, you need to think through what you're going to do. You need to locate every wall precisely on the subfloor, along with every framing member in those walls.

Layout Starts in the Office

For one of our typical houses, layout and framing for interior and exterior walls start in the office a few days before my crew and I are ready to pick up the first 2×6. First, I review the plans carefully and make sure that all the necessary information is there.

I need the locations and dimensions of all the rough openings, not only for doors and windows but also for things such as fireplaces, medicine cabinets,

LAYOUT IS KEY. Everything from rafters to kitchen cabinets fits better when you get the walls square and the studs in the right place.

built-ins, dumbwaiters, and the like. I also make sure the plans have the structural information I need for layout, such as shear-wall and bearing-wall details and column sizes.

At the site, one of Spier's many corollaries to Murphy's Law is that errors never cancel each other out; they always multiply. If the floor is anything but straight, level, flat, and square, the walls are going to go downhill (or uphill) from there. So before you get to layout, do whatever it takes to get a good floor, especially the first: Mud the sills, shim the rims, rip the joists. Sweep off the subflooring, and avoid the temptation to have a pile of material delivered onto it.

Snap Chalklines for the Longest Exterior Walls First

I've learned over the years that it's best to snap the plate lines for the entire floor plan before building any of it. Problems you didn't catch on the prints often jump out when you start snapping lines.

When framing floors, I take great care to set the mudsills flush, square, and in their exact locations. Because the edges of floor framing and subflooring are not always perfect, though, I use a level to plumb up from the mudsills and establish the plate lines, measuring in the stock thickness from the level. I generally start with the longest exterior walls and the largest rectangle in the plan. When I have the ends of the longest wall located, I snap a line through the marks.

Once I've established the line for the first wall, I move to the parallel wall on the opposite side of the house. I measure across the floor from the first line to the opposite mudsill (again using a level to plumb up from the mudsill to the floor height) at both ends; if the lengths differ slightly, I use the larger measurement. I snap through these points, which gives me two parallel lines representing the long sides of the largest rectangle. It's okay if the plates overhang the floor framing by a bit, but I watch for areas that might need to be shimmed or padded—for instance, where a deck ledger needs to be attached to the house.

TAKE LAYOUT LINES FROM THE MUDSILLS

If the mudsills were installed perfectly square, you can avoid any discrepancies in the deck framing by plumbing up from the mudsills and measuring in from there.

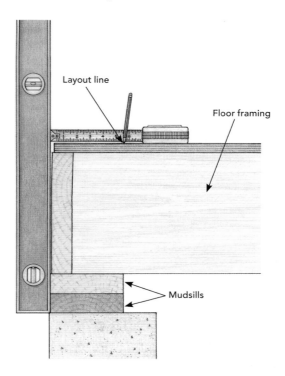

Layout line

Floor framing

Mudsills

Establish the Right Angles

I locate three corners by measuring in from the mudsills. The fourth corner I locate by duplicating the measurement between the first and second because I need sides of equal lengths to create a rectangle. I check this rectangle for square by measuring both its diagonals. If I've done everything right so far, the diagonal measurements should be very close, perhaps within ¼ in. I shift two corners slightly if I need to, making sure to keep the lengths of the sides exact until the diagonals are equal. Now, perpendicular lines are snapped through the corners, completing the rectangle.

Because I started arbitrarily with one long wall, I may find now that the rectangle, although perfectly square, is slightly askew from the foundation and floor. Also, some of the complicated foundations

SETTING THE STAGE FOR THE REST OF THE HOUSE

FOR THE MOST PRECISE WALL layout, plot a series of rectangles that includes every wall. The larger the rectangle, the more accurate the wall position. Begin with the longest walls, and lay out the largest rectangle using diagonal measurements (see the photo at right). Working off established lines and square corners, work down to the smallest rectangle.

EQUAL DIAGONAL MEASUREMENTS MEAN A SQUARE LAYOUT. After snapping chalklines for the longest parallel walls, the author takes corner-to-corner measurements to make sure the corners are square for a perfect rectangle.

1. Starting with the longest walls, measure and square the largest rectangle.

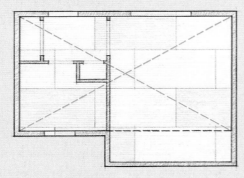

2. Working off those lines, plot the rectangle that includes the jog in the wall.

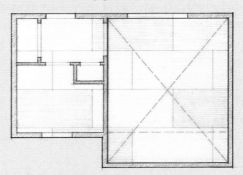

3. Now measure off the outside and form a rectangle for the longest interior wall.

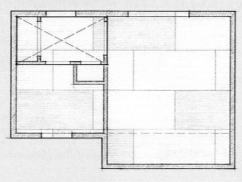

4. Last, form rectangles for the remaining interior walls.

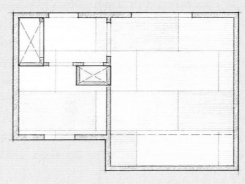

It's a rare architect who dimensions a plan to the fraction of an inch with no discrepancies, and an even rarer builder who achieves that accuracy. So first I lay out critical areas such as hallways, stairwells, chimneys, and tub or shower units, and then I fudge the rest if I need to.

that I work on can have wings or jogs that are slightly off. If I can make everything fit better by rotating the rectangle slightly, I take the time to do it now.

Smaller Rectangles Complete the Wall Layout

With the largest part of the plan established, I lay out and snap whatever bays, wings, and jogs remain for the exterior walls. I use a series of overlapping and adjacent rectangles, which I can square by keeping them parallel to the lines of the original rectangle. I again check the right angles by measuring the diagonals.

Often, the plan calls for an angled component such as a bay. If these components are at 45 degrees, I lay them out from right angles by forming and diagonally bisecting a square. For angles that are not 45 degrees, I either can trust the architect's measurements on the plans, or I can use geometry and a calculator. The latter method is more likely to be accurate.

When all the exterior walls have been laid out, I turn my attention to the interior walls. Again, I start with the longest walls and work to the smallest, snapping lines parallel and square to the established lines of the exterior walls. I snap only one side of each plate, but I mark the floor with an X here and there to avoid confusion about where the walls will land. I also write notes on the floor to indicate doors, rooms, fixtures, bearing walls, and other critical information.

One last critical issue when reviewing the plans and laying out walls is watching for elements of the design that need to stay symmetrical. If the foundation contractor made one wing a bit wider than another, you don't want to build all three floors before realizing that the ridgelines of the two wings needed to match up. Make sure symmetrical elements are aligned at the first layout stage.

Make Plates from the Straightest Lumber

While I snap the walls, the crew is busy cutting and preparing material from the piles of stock. I have them set aside a pile of the straightest lumber. With the chalklines all snapped and with this material in hand, I start cutting and laying out the plates (top and bottom members) for the exterior walls. In this

CHALKLINE TIP. To snap chalklines for short walls, hook one end of the line to your boot and stretch the other end out to the mark. Rotating your foot slightly aligns the boot end, and you're ready to snap.

PLATES ON DECK. When wall positions are laid out, cut all the top and bottom plates (the long horizontal members) for the exterior walls and place them on their layout lines.

step of layout, we set the plates side by side on the layout line, and every wall-framing member is located and labeled. With this information, we assemble the walls on the floor, then raise them into place. I often call out measurements and have someone cut and hand up the material to keep mud, snow, and sawdust off the floor during this crucial phase.

As a rule, we plate the longest exterior walls to the corners of the house, and the shorter walls inside them. This approach sometimes needs to be modified—for instance, to accommodate structural columns, hold-down bolts, or openings adjacent to corners. Sometimes an obstruction or a previously raised wall dictates which wall can be built and raised first. The goal here is to build and raise as many walls as possible in their exact positions, especially the heavier ones. Moving walls after they're raised is extra work and no fun.

Before starting any framing, I established a common-stud layout for the entire structure based on two long perpendicular walls from which layout for the rest of the house framing can be measured. This common layout keeps joists, studs, cripples, and rafters throughout the house vertically aligned from the foundation to the ridge, which makes for

a strong, straight, and easily finished structure. We use this common layout to locate butt joints between pairs of plates because code and common sense dictate that these joints land on a stud or a header.

As my crew and I measure and cut the pairs of wall plates, we lay them on edge along their layout lines, sometimes tacking them together with just a few 8d nails to keep the plates held together and in place.

Window, Door, and Stud Layout at Last

When all the exterior plates are in place, it's finally time to lay out the actual framing members. I always start with the rough openings for windows and doors. Most plans specify these openings as being a measured distance from the building corner to the center of the opening, which works fine. You can allow for the sheathing thickness or not, but once you choose, be consistent, especially if openings such as windows have to align vertically from floor to floor. Obviously, if an opening such as a bay window or a front door is to be centered on a wall, center it using the actual dimensions of the building, which may differ slightly from the plan.

Rough openings are a subject worthy of their own chapter, but in a nutshell, I measure half the width of the opening in both directions from the center mark. I then use a triangular square to mark the locations of the edges of trimmers and king studs, still working from the inside of the opening out. Various other marks, such as Xs or Ts, identify the specific members and their positions.

Next, I mark where any interior-wall partitions intersect the exterior wall. At this point, I just mark and label the locations; I decide how to frame for them later. I also locate and mark any columns, posts, or nailers that need to go in the wall. I lay out any studs that have to go in specific locations for shelf cleats, brackets, medicine cabinets, shower valves, cabinetry, ductwork, and anything else I can think of. Doing this layout now is much easier than adding or moving studs later.

STUD LAYOUT IS ALWAYS TAKEN FROM THE SAME TWO WALLS. One crew member holds the tape at wall offset while the other marks the stud position (above). Even when there is a break in the wall, the layout is pulled from the same place to keep all the framing aligned (left).

LAYING OUT MULTIPLES. For things such as short closet wall plates, line them up and draw two walls' worth of layout lines at once.

Finally, I lay out the common studs on the plates. Studs are commonly spaced either 16 in. or 24 in. on center to accommodate standard building products. By doing the common-stud layout last, I often can save lumber by using a common stud as part of a partition nailer. I almost never skip a stud because it's close to another framing member, which, I've learned the hard way, almost always causes more work than it saves. I occasionally shift stud or nailer locations to eliminate small gaps and unnecessary pieces. I keep the plywood layout in mind here, though, so that I can use full sheets of sheathing as much as possible.

Inside Walls Go More Quickly

Once the exterior walls are built and standing, I cut the interior-wall plates and set them in place. Where two walls meet, I decide which one will run long to form the corner so that the walls can be built and raised without being moved. Also, facing a corner in a particular direction often provides better backing for interior finishes, such as handrails or cabinetry, and sometimes is necessary to accommodate such things as doorways or multiple-gang switches.

When the plates are cut and set in place, I do the stud layout. Just as with the exterior walls, I do the openings first, then nailers and specific stud and

COPY THE LAYOUT FROM THE PLATE. To mark the cripple layout on the rough windowsill, just line it up on the plate and copy the layout.

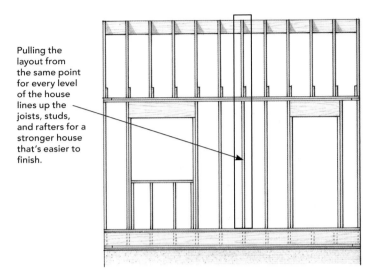

Pulling the layout from the same point for every level of the house lines up the joists, studs, and rafters for a stronger house that's easier to finish.

column locations. Next, I mark the locations of intersecting walls and finally overlay the common-stud layout on the plates.

Where Walls Come Together

Where one wall meets the middle of another, I use a partition post if the situation dictates it, but more often, I opt for an L-nailer. To make an L-nailer, I use a wider stud on the flat next to a common stud whenever possible. It's faster and easier; it accommodates more insulation; and it saves the subs from drilling through those extra studs and nails. If I use U-shaped partition posts (a stud or blocks on the flat flanked by two other studs) in an exterior wall, I need to make sure to fill the void created by the partition post with insulation before the sheathing goes on.

With the interior plates all there, we can nail in the studs, raising walls as we go. I mark key areas where studs should be crowned or specially selected, such as areas with long runs of cabinetry, and also studs that might need to be left out to allow installation of things too wide to be carried though the doors. I also nail double top plates to as many walls as possible if they don't interfere with the lifting process.

FRAMER'S SHORTHAND: WHAT THOSE LITTLE MARKS MEAN

WHEN THE FRAMING members are marked, a full-length stud is indicated by an X. A trimmer or jack stud is a T or J, and a C or X indicates a cripple (a short framing member below a sill or above a nonstructural header). Other framing, such as partition posts and corner posts, are labeled, along with any special framing instructions.

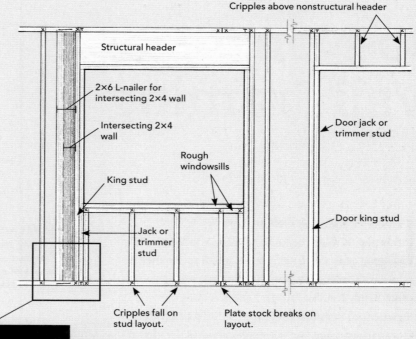

Cripples above nonstructural header

Structural header

2×6 L-nailer for intersecting 2×4 wall

Intersecting 2×4 wall

Rough windowsills

King stud

Door jack or trimmer stud

Door king stud

Jack or trimmer stud

Cripples fall on stud layout.

Plate stock breaks on layout.

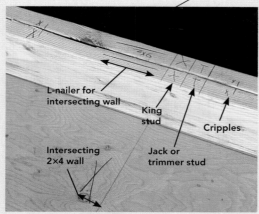

L-nailer for intersecting wall

King stud

Cripples

Intersecting 2×4 wall

Jack or trimmer stud

SPELL OUT WHAT YOU NEED ON THE WOOD. Mark key areas where studs might need to be left out to allow installation of things too wide to be carried though the doors, such as one-piece tub/shower units.

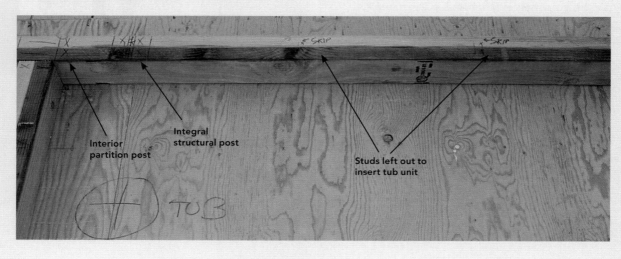

Interior partition post

Integral structural post

Studs left out to insert tub unit

No-Nonsense Wall Framing

BY MIKE NORTON

If there's a glamorous job in carpentry, it's not framing. It might be finish work; everything looks so good after that final piece of molding is nailed in place and the job is complete. Framing, on the other hand, is called "rough," and it requires an experienced imagination to see the finished product in its earliest stage. But framing embodies the physics of the structure, and if you don't get it right, the house will fail. You'll also have a hard time nailing your fancy trim where there is no blocking.

Whether the plans call for traditional stick framing or optimum-value engineering, the skills are relatively simple: straight and square cuts, a good hammer technique, economy of motion, and a strong back. There are a few tricks, however, that make the job easier and the results more professional. Here, I'll explain some of the methods I've picked up, using a simple exterior wall as an example.

Layout Is Critical

Even though you're probably the one who laid the sill plates and framed the deck, it's still a good idea to make sure that the deck is square before framing the walls. The simplest method to use is to check the corners by measuring a 3/4/5 triangle and then extending the angle with a reference chalkline. I've

GET SQUARE AND PARALLEL WALLS WITH A LASER. A key step in laying out walls is to establish square corners and parallel walls, a task streamlined with the use of a laser. Once square corners have been set, the plate layout lines can be snapped.

VERIFY THE 90. A check of the target placed on the opposite corner verifies the 90-degree angle generated by the laser. Adjust the plate location if the original mark doesn't match the target.

A GOOD LASER MAKES LAYOUT FASTER AND MORE ACCURATE. The author uses a laser that fires a constant beam in five directions that is visible in daylight (Pacific Laser Systems PLS-5 kit). The first step is to mark the plate width in all corners, set the laser, and aim it toward the first target.

HIT THE BULL'S-EYE. Placed on the next corner over the plate mark, the target indicates when the laser is lined up.

ESTABLISH PARALLEL LINES. From the corners verified by the laser, measure and mark the opposite wall-plate positions (4). Snap layout chalklines for the plates (5).

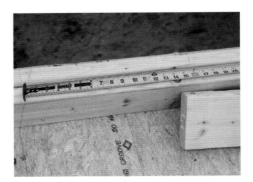

LAY OUT THE PLATES ONCE. It's usually easiest to build walls on the deck and then lift them into place. The first step is to mark out stud, window, and door openings on the plates. A length of metal packing strap (shown) nailed to the plate and deck is good insurance against a wall dropping off the deck's edge as it's raised.

EFFICIENT LAYOUT. For a quick layout, place the top plate behind the bottom plate and mark both at the same time. This temporary placement also keeps deck clutter to a minimum.

TOE THE PLATE. Lay the bottom plate on edge and align it with the chalkline. Toenail it to the deck, nailing every 2 ft. or so. Once the wall is built, the completed structure is anchored to the deck and is less likely to move as it's raised.

found it faster, however, to use a laser that shoots two lines at 90 degrees. Two people can square up a deck in about 10 minutes, and there's less chance for error. If the deck isn't square, it's usually within a ¼-in. tolerance that we can correct by moving the plate location marks out beyond the deck edge or inward toward the center of the deck.

After squaring up the deck, we snap lines for the plates. While we're at it, we also snap a reference centerline across the deck so that we can check that walls are parallel or, when it's snapped to represent the ridgeline, use it to lay out a gable wall.

We frame the walls flat on the deck by first toe-nailing the bottom plates down on edge along the chalklines. After double-checking the window and door schedule, we mark these locations on the plates. I also figure out where the partitions intersect the wall and mark the location of the backers. If there's a conflict between the partition's placement and the eventual locations of interior trim, I usually call the architect before making the necessary adjustments.

Framing layout is a critical part of the process, so I always double-check my measurements. I cut the bottom plate to length first, then the top plates. I usually wait to install the second top plate until adjoining walls are raised so that the plate ties the walls together.

With the bottom and top plates placed together temporarily, I start marking the layout from the left and go right. After I mark the first stud on the bottom plate, I drive a nail at the line and pull 16 in. from there, marking the X beyond the line that indicates the stud location. At the same time, I transfer the layout to the top plate.

If I have a straight wall and a simple floor frame above, I mark the floor-joist layout onto the second top plate's face so that we don't have to do it after we lift the wall.

AN ORGANIZED LIST. After assigning each window a letter designation and listing the parts' measurements, cut and stack the parts in discrete piles.

PRODUCTION-STYLE ASSEMBLY. Assigning one carpenter the task of cutting parts at the chopsaw station means that the rest of the crew can keep nailing. If there are a number of identical windows, it's faster for the cut man to make parts and assemble the windows as units.

ALWAYS CHECK FOR SQUARE. After the wall has been nailed together, measure both diagonals to make sure the wall is square before starting to sheathe.

EASIER ON THE GROUND. It's safer, faster, and neater to cut out the window openings with a circular saw as the sheathing is installed.

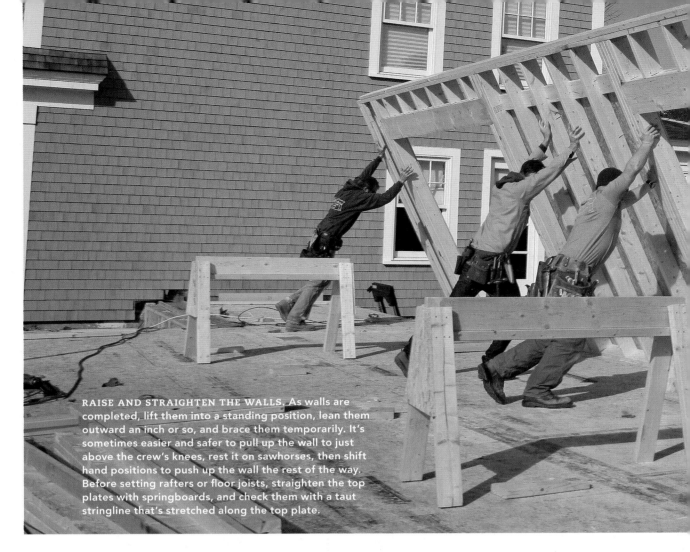

RAISE AND STRAIGHTEN THE WALLS. As walls are completed, lift them into a standing position, lean them outward an inch or so, and brace them temporarily. It's sometimes easier and safer to pull up the wall to just above the crew's knees, rest it on sawhorses, then shift hand positions to push up the wall the rest of the way. Before setting rafters or floor joists, straighten the top plates with springboards, and check them with a taut stringline that's stretched along the top plate.

SET THE SPRINGBOARDS. With the wall braced, start the straightening process by nailing one end of a pine 1×8 to the underside of the top plate or to a header.

TACK THE BOTTOM. While one carpenter pushes down on the board, the other checks the string with a 2× gauge. The trick is to increase the tension by overbending the board so that it pulls the wall into the string. The first carpenter then tacks the springboard end to the deck and releases the tension on the board.

PULL THE WALL. One carpenter pushes a shorter leg into the underside of the springboard, bringing the top plate back toward the string. When the gauge indicates that the top plate is straight, the other carpenter nails the leg to the springboard, locking it and the wall into place.

TRICK OF THE TRADE. Use this foolproof nonslip knot to stretch the stringline taut when straightening walls. To start pull the string tight, make a loop around your finger, and twist it.

NEXT. Anchor the loop on the lower nail, and pull up on the string.

Cutting Duplicate Components All at Once Is Faster

I've found that it's more efficient to have one of the crew designated as the cut man at a chopsaw. (Mounted on a stand with adjustable stops, the saw makes production work simple and accurate.) Wall studs, headers, window and door parts, and other duplicates all get cut there.

If the house has many of the same windows, I put together a cut list so that we can cut the legs below the windows and the cripples above at the same time as the headers and sills. On the list, I group the windows with the same header length so that the cut man can cut everything without having to adjust his jig.

It's usually easier for the cut man to cut and assemble the headers, jacks, sills, legs, and other pieces into door and window units that we then can incorporate into the frame. If they're all various sizes, we assemble the doors or the windows as part of the wall. When the house is fairly complicated and has different wall heights and large windows of different sizes, I have someone ready on the deck to cut all

FINALLY. While keeping tension on the string, pull down hard to compress the wraps on the loop. Loosen the tension on the string, and the knot comes undone.

the legs and cripples as we frame the wall. Because the width of the header stock can vary by ¼ in. from one 2×10 to another, most times we cut the cripples to length, install them, and then cut the jacks to the corresponding length. The rough opening's height may decrease by a fraction of an inch, but as long as the window fits in the rough opening, it's OK.

Sheathing Layout Is Important

Once we've assembled a wall on the deck and measured its diagonals to make sure it's square, we begin sheathing. We pay careful attention to the engineer's plans, including the nailing pattern on the sheathing and the location of vertical sheets of plywood to hold down the corners of the house. If a wall is over a certain height or the engineer requires longer sheets for a hold-down, we install blocking across the wall at the point where the seams meet.

We often use 4×8 plywood sheets to span from the mudsill to about a foot under the top plate of the wall. We then can use 4×10 sheets to span from that point to the second top plate of the second-floor wall. We also cut out windows and doors as we sheathe the walls, rather than doing so after we stand them up. It's safer, faster, and more precise.

Bracing and Straightening

After we've raised a wall, we nail the plates to the deck and temporarily brace the wall with 2×6s so that it's pushed outward slightly. This makes it easier to raise an adjacent wall; it's also easier to pull the wall straight than to push it.

To straighten the walls, we nail a 2× block to each end of a wall at the top plate, then run a taut line between the two. Next, we install roughsawn 1×8 springboards at 8-ft. intervals. We cut them to length so they can span a 45-degree angle from the top plate to the deck, then we nail one end to the underside of the top plate. While one person checks the string with a 2× block, another flexes the board downward and tacks the lower end to the deck.

We jam a short length of board between the deck and the middle of the springboard and push it away from the wall until the gauge block shows that the wall is straight. Then we tack the short board to the springboard and the deck, locking the wall in.

Laying Out and Detailing Wall Plates

BY LARRY HAUN

As a beginning carpenter, I was so afraid of making a mistake when laying out walls that the process took me hours. Layout seemed like an exact science, one whose rules I didn't know.

Transferring measurements from blueprints to full-size markings on the floor and then cutting the wall plates to fit these markings sets the stage for all of the carpentry that follows. If the walls are out of parallel or if rooms are not square, roof rafters, cabinets, and even floor tiles won't fit properly.

Done correctly, though, wall layout leaves a set of clearly marked plates—templates, in a sense, that can enable relatively inexperienced carpenters to assemble the walls. Well-laid-out walls compensate for errors in the floor, be it slab or frame, and ensure that fixtures such as bathtubs fit between the walls. Once I learned the rules, learned where being off ¼ in. matters and where it doesn't, I relaxed and then began to enjoy laying out walls.

The Plans May Not Tell the Whole Story

Before beginning to frame a house, I study the plans at home. I note what the plans show and what must be inferred. There are often differences between plan dimensions and what is needed on the job. For exam-ple, plans may call for 36 in. between walls for a set of stairs. So that the stairs themselves will measure a full 36 in., I leave at least 38½ in. from stud wall to stud wall. The extra 2½ in. leaves enough room for ½-in. drywall and a ¾-in. skirtboard on each wall, enclosing 36-in. treads and risers between them.

To accommodate a bathtub, most bathrooms need to be 60 in. wide. I usually frame bathrooms ⅛ in. or ¼ in. big so that the tub can be installed with ease. Nonstandard-size tubs are becoming more common, and in these cases the tub supplier may be a more re-liable source for the tub's dimensions than the plans.

Plans often call for halls to have 36 in. between fin-ished walls. To achieve this dimension, the studs and plates must be 37 in. apart, allowing for ½-in. drywall on each wall. For that matter, 36-in. halls, common on older stock plans, don't allow adequate room for code-required 2/8 (32-in.) doors and their trim.

I lay out halls ending at a 2/8 door with 40 in. between stud walls, which leaves room for a 37-in. header with a king stud on each side. Typically, the extra inches are stolen from bedrooms on both sides of the walls, a theft that is hard to notice.

I check that all the plan dimensions are correct. It's not uncommon for the room dimensions and the thicknesses of their walls to add up to a different number than that called out as the overall size of the

KING AND JACKS ARE MORE THAN A POKER HAND

They hold up your house as well. Through walls are continuous parallel walls that run the house's long dimension. Butt walls fit between them. Cripples, studs, headers, plates, and channels comprise both types of wall. The drawing below is a glossary of these builder's terms.

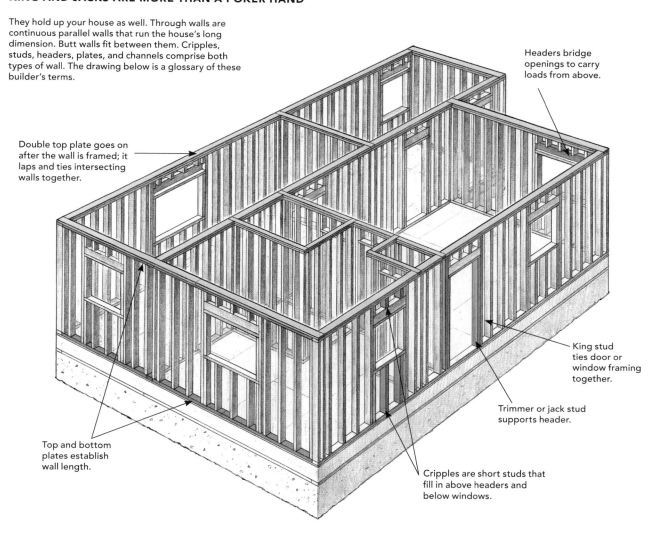

Headers bridge openings to carry loads from above.

Double top plate goes on after the wall is framed; it laps and ties intersecting walls together.

King stud ties door or window framing together.

Trimmer or jack stud supports header.

Top and bottom plates establish wall length.

Cripples are short studs that fill in above headers and below windows.

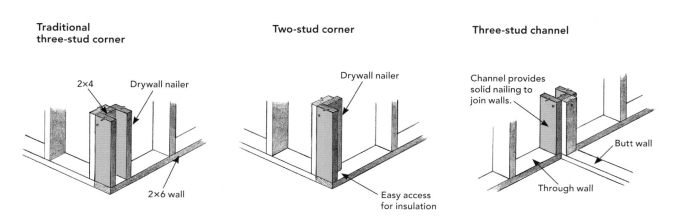

Traditional three-stud corner

2×4

Drywall nailer

2×6 wall

Two-stud corner

Drywall nailer

Easy access for insulation

Three-stud channel

Channel provides solid nailing to join walls.

Butt wall

Through wall

PLANS MAY CALL FOR A 36-IN. STAIRWAY, BUT THAT OFTEN MEANS 38½ IN. BETWEEN THE STUDS

Stairs are often site-built and the framer must leave space for drywall and skirtboards, as well as for 36-in. treads.

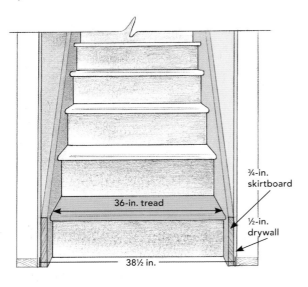

¾-in. skirtboard

36-in. tread

½-in. drywall

38½ in.

CODE SPECIFIED 2/8 DOORS DON'T FIT WELL IN 36-IN. HALLWAYS

37-in. header for 2/8 door

The author often widens halls to 40 in., adding room to trim around the door.

Codes now specify 2/8 doors into most rooms. Stock plans often show only the enlarged door.

Hall widened to 40 in.

Fitting a 2/8 door in a 36-in. hall requires insetting the kings and jacks in the through walls and ripping the casing.

Plate   Drywall

36 in.

Jamb and casing

Jack   King

Enlarging the hall to 40 in. provides room for both the framing and full trim.

Plate   Drywall

40 in.

Jamb and casing

Jack   King

house. Figuring out these discrepancies is much easier at my kitchen table than in the field. I note which walls are 2×4 (most interior walls) or 2×6 (most exterior walls and walls with plumbing). I make notes directly on the plans so that all the needed information is in one place, and I discuss any wall-location changes with the owner, builder, or architect before snapping the wall lines.

Although following plan dimensions is important, most builders view plans as a guide and not as law written in stone. For example, when working on a concrete slab, you might find that a pipe will be 1 in. or so off layout. Fixing this error would be a huge job for the plumber, and it would slow the builder's schedule. Rather than snap the wall line to the plan dimension, you can usually move the wall to cover the pipe. Check first, though, that moving this wall doesn't affect adjoining areas where space is crucial.

Check Floors for Parallel and Square

Once you're familiar with the plans, it's time to sweep the floor and to begin laying out walls. The first step always is to check that the floor is parallel and square. I measure the floor from outside to outside at both ends of the longest exterior walls. If the measurements are equal, I mark the inside of the walls on the floor using a scrap of 2×4 or 2×6 plate stock as my guide. Pencil shows up well on wood, but I use keel, or lumber crayon, on concrete.

From these marks, I snap chalklines that establish the insides of the long walls. An awl driven into a wood floor holds the end of a tape or chalkline. I use a weight on concrete slabs; in green concrete, you can drive a nail into the slab.

Tweaking Walls Square on an Out-of-Square Deck

If the exterior walls are slightly out of parallel, adjustments can be made. The last house I helped to frame was built on a slab whose long sides were 1 in. out of parallel. I moved both walls inward ¼ in.

TIP

Be sure to stretch chalklines tight and to pull the line straight up when snapping. Chalklines released at an angle can leave a curved line on the floor. Snapping lines in the wind can also leave a curved line.

at the wide end of the slab, decreasing the width ½ in. At the narrow end, I moved each wall out ¼ in., increasing the width ½ in. and making the walls parallel. Sometimes nothing you can do will get the walls perfectly parallel. If I can get them to within ¼ in. over 12 ft., that's usually good enough.

Keep in mind what will cover the exterior walls. Wall coverings often extend below the framing onto the slab or foundation by 1 in. or so. If you move a wall in on the foundation too far, the siding may not be able to extend below the sill.

ADJUST WALL TO FIT PLUMBING

Sometimes the plumber gets it wrong. Moving walls a few inches often puts pipes in the right spot.

Right

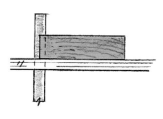

Wrong

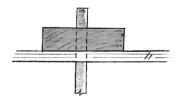

SQUARING PLATES ON AN OUT-OF-PARALLEL SLAB

Check first that the two longest sides of the deck or slab are parallel. If they aren't, you'll need to adjust them, and snap the chalklines, marking these plates parallel.

← Chalklines

With the long sides parallel, square the perpendicular outside walls. Remember, any triangle whose sides measure 3, 4, and 5, or multiples thereof, is a right triangle.

A

B

10 ft.

8 ft.

6 ft.

Section A

Plates can be inset by the thickness of the sheathing, or more with some sidings, to square up walls.

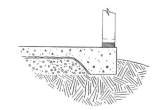

Section B

Plates can hang over the deck as much as 1 in.

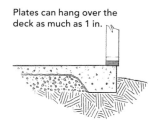

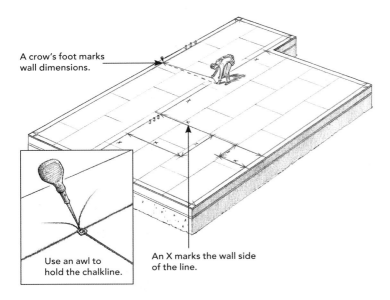

A crow's foot marks
wall dimensions.

Use an awl to
hold the chalkline.

An X marks the wall side
of the line.

Snapping lines for the perpendicular exterior walls comes next. I mark the wall location on the floor at each end of the first perpendicular wall, again using plate scrap as a guide. After connecting these marks with a chalkline, I check the corner for square by measuring 3 ft. along one wall and 4 ft. along the other; then I check this triangle's hypotenuse. It will measure 5 ft. if the corner is square. Multiples of 3, 4, and 5 work as well, so a 6-8-10 triangle is also a right triangle.

If a corner is slightly out of square, you can adjust it much as with parallel walls. Move one end of the wall out a bit and the other end in. Take special care to make sure the kitchen and bathrooms are square so that cabinets and tile can be set easily.

Measure from the Exterior Walls to Lay Out the Interior Walls

Measuring from the long exterior walls, I can now mark the parallel walls on the floor. A plan dimension might read 12 ft. 6 in., for example, outside to center of a 2×4 wall. To locate the edge of a wall, simply move the mark over 1¾ in., or half the thickness of a 2×4. I locate both ends of an interior wall

by measuring from an exterior wall and marking a V or crow's foot on the floor. Long walls such as those for bedrooms and hallways are laid out before the short walls for closets and bathrooms are located.

I always make an X with keel alongside the crow's foot to show on which side of the mark the plate will fall. Be careful: An X on the wrong side of a bathroom wall, for example, may mean that the tub won't fit. Some carpenters snap two chalklines to note that the plate will fall between the lines. This seems to be an unnecessary extra step.

All of the chalklines should be snapped straight through wall openings. Pay no attention to door and window openings when snapping chalklines.

I note anything out of the ordinary on the floor with keel. For example, I indicate the end of a short wall with a mark and write "end" at that point. If a plumbing wall is to be 2×6, I write that on the floor. I try to keep it simple because too many marks can be just as confusing as too few.

Distribute All the Plate Stock before Cutting Any to Length

After I finish snapping all the lines, the next step is carrying and placing two pieces of plate stock along every wall line. These pieces of stock will be the top and bottom plates, and their lengths should approximate the wall length. To ensure straight and strong walls, I use long (16 ft., if possible), straight stock for plates. It's a good idea to sight the plate stock and to use the straighter pieces for the top plates. The bottom plates can be easily straightened when they're nailed to the floor, but the top plates must be braced straight after the walls are raised.

Once the plate stock is distributed, I start plating the longest outside walls. These plates run through from corner to corner and are called through walls. Walls that fit between other walls are called butt walls. The chalklines mark the exterior walls' inside edge. Because the chalklines represent the inside of the walls, I locate the exact ends of the first through walls by lining up the end of the plates on a scrap of second, or double, plate will be nailed to the top plate, reinforcing it and tying together the walls. At least 4 ft. must separate joints in these top plates. So that intersecting walls can attach solidly, I keep joints in both top plates 4 ft. from intersections.

Once the exterior walls are done, I cut, stack, and tack (with 8d nails) the plates for the interior walls that parallel the long exterior walls. All perpendicular walls will butt into these through walls. I run all plates continuously, ignoring door and window openings. I cut the bottom plate for door openings later.

Cut the Plates to Length by Eye

The wall lines snapped on the floor show the position and length of each plate. It takes a bit of courage and practice, but you should be able to cut wall plates square and to length without further use of measuring tape or square. For through walls, sight the circular-saw blade on the chalkline below. Line up cuts on butt-wall plates with the intersecting through walls. If you have a 10-in. saw, you can tack the plates together and cut them with one pass.

The bottom plates can be cut a bit short. The result will be a harmless gap where two bottom plates

plate stock held on the intersecting wall's chalkline and cut the plate to length.

I tack the bottom plate to the floor with two or three 8d nails, stack on the top plate, and tack it to the bottom plate. I continue stacking and tacking until I reach the far end of the wall. There, I mark the ends of the plates as I began, with a scrap held to the butt wall's chalkline, and cut the plates to length.

Staggering joints in the top and bottom plates is okay, but not necessary. The bottom plate will be nailed to the floor. After the studs are nailed in, a

PLATE THE LONGEST WALLS THROUGH FROM END TO END

Plate all walls parallel to the long exterior walls as through walls. All other walls butt to these. This eases raising the completed assemblies.

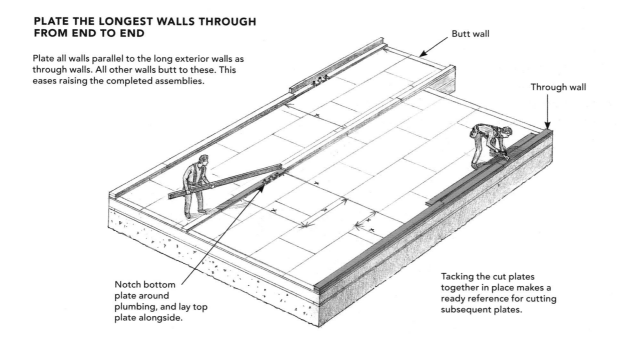

Butt wall

Through wall

Notch bottom plate around plumbing, and lay top plate alongside.

Tacking the cut plates together in place makes a ready reference for cutting subsequent plates.

QUICKLY CUT THROUGH WALLS. Using a circular saw, line up cuts on butt-wall plates with the intersecting through walls.

ANCHOR-BOLT MARKER. Lay the plate exactly on the opposite side of its chalkline, center the bolt in the marker, and then whack it with a hammer to mark where to drill the plate. The second bolt is for 2×6 plates.

intersect. The top plates, though, must be cut within 1/16 in. Otherwise they'll push or pull walls out of plumb.

Plating on a concrete slab is much the same as on a wooden deck, but there are a few differences. Remember to use treated wood for the bottom plate. The bottom plate can't be as easily tacked to a slab as it can be to a wood deck. I simply lay the bottom plates on their lines and tack on the top plates. I toenail intersecting plates together so that they don't move until I'm ready to nail together the walls.

Often, the bottom plate of an exterior wall must be bolted fast to a concrete slab. Typically, anchor bolts are cast into the edge of the slab, and holes are drilled in the plate to receive the bolts. I mark the holes with a bolt marker. Set the plate directly on the chalkline, but on the opposite side of where it will finally go, and mark the hole location. Interior walls on slabs are usually nailed home with powder-driven pins or with concrete nails. Codes are specific about what fasteners can be used, so if in doubt, ask your building department.

When plumbing is roughed in, as it is on a slab, plates can't be stacked. I notch the bottom plate to fit around the pipes, lay it on the line and put the top plate next to the bottom.

Mark Every Stud on the Plates

Detailing plates is marking the location of all corners and intersecting walls, headers, and studs in every wall on the plates. Because hundreds of marks are made on the plates, I keep this marking system simple. Extra marks are confusing.

I start detailing by marking the locations of corners and channels (three-stud assemblies in through walls to which butt walls are nailed). These spots require extra studs so that walls can be nailed together properly. The extra studs also provide backing for drywall inside and sheathing outside.

I lay out corners and channels by lining up a channel marker on the edge of the intersecting wall and drawing a line on the top and both sides of the through wall. I mark an X with keel on the top plate to show the framers where the corners and channels will nail in place. Mark corner and channel locations precisely. A repeated small mistake can cause walls to be out of plumb once they are raised. The corner and channel marks also indicate where the double top plates will intersect, tying together the through walls and butt walls.

Next, I mark the location of each door and window with keel. I will have made all the headers early. I lay them at their place on the plates, and then simply draw a keel line down from the header and across the two plates without using a square. Away from the header and on each plate, I mark an X at the location of the king studs that nail to the header ends. A straight line on each plate under the header indicates this space is a door or window opening. I also mark the lengths of upper and lower cripples on the headers.

At this point, I study the plans again and detail any specials such as medicine cabinets, tub backing, openings for plumbing access, posts for beams, heating ducts, short or tall walls, or even ironing boards.

Proper layout for these items before framing means you won't have to remodel a wall later.

Use a Layout Stick to Mark the Studs

The framing members in most houses are laid out on 16-in. or 24-in. centers because these measurements divide evenly into the 4×8 dimensions of standard sheathing and drywall. Maintaining a consistent layout along the length of the exterior wall is important so that panel edges always fall in the middle of a stud.

WELL-MARKED PLATES MEAN CARPENTERS DON'T NEED TO THINK

After all the plates are cut, the author sets the headers in their places and marks the location of every stud, channel, and corner on the plates. When finished, the plates provide a framing map that even inexperienced carpenters can follow.

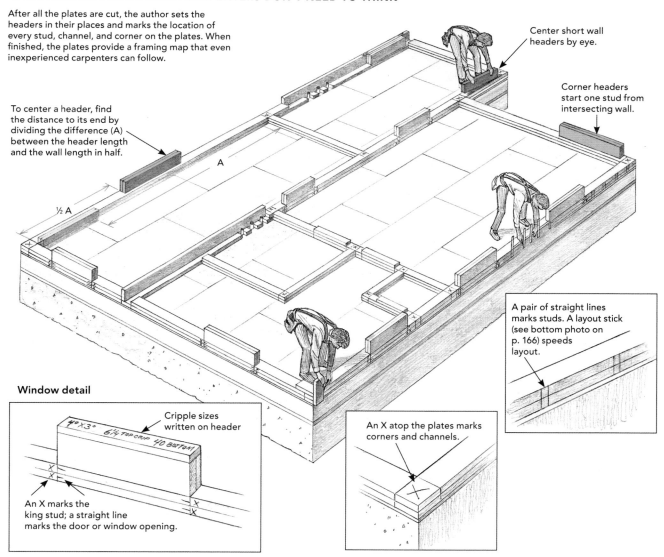

To center a header, find the distance to its end by dividing the difference (A) between the header length and the wall length in half.

½ A

A

Center short wall headers by eye.

Corner headers start one stud from intersecting wall.

A pair of straight lines marks studs. A layout stick (see bottom photo on p. 166) speeds layout.

Window detail

Cripple sizes written on header

An X marks the king stud; a straight line marks the door or window opening.

An X atop the plates marks corners and channels.

CHANNEL MARKER. This lightweight 3½-in.-wide tool enables carpenters to mark intersecting butt walls on all three sides of through walls with a pencil. A similar tool can be quickly site-built of plate scrap.

Many carpenters lay out stud locations by stretching out a tape measure, ticking the 16-in. or 24-in. increments on one plate with a pencil and returning with a square to extend the lines to both plates. This process works, but it's slow.

I use a layout stick instead. This 4-ft. long aluminum bar has 1½-in. wide tabs on 16-in. and 24-in. centers and serves as a template for spacing studs. To begin a wall, I hang the first short tab on the stick ¾ in. beyond the end of the wall. This placement sets up the layout so that the centers of the studs fall on 16-in. or 24-in. centers, and sheathing edges will land centered on the studs. Then I mark both sides of all the studs for this 4-ft. section of wall, move the layout stick down, line its end up on my last mark, and again mark out 4 ft. more of wall.

When I come to a door or window opening, I continue the layout, marking cripple locations on the headers. You can ease the plumber's task by laying out the studs to leave a full bay for shower and bath valves. I like to start butt-wall layout at channels,

leaving a full bay's room to swing a hammer when nailing the walls together.

Because of the sheer number of marks that are required, I wander through the plated rooms and check to see whether I have missed marking a corner, a door, or even some studs here and there. An error caught at this point can save time and grief during the actual framing.

At this point, all the information necessary to frame the walls is marked on the plates. Even if I have to leave the site, a relatively inexperienced crew can have the walls standing by the end of the day without ever seeing the plans. The final step in plating comes after the studs have been nailed between the top and bottom plates. I nail on the double top plate before raising the walls. On through walls, I leave out sections of this plate at the corners and channel marks. The double top plates on the adjoining butt walls are cut one plate width longer to lap the through wall above corners and channels, tying the walls together.

LAYOUT STICK. With tabs on 16-in. and 24-in. centers, a layout stick locates studs fast. The longer teeth come into play when plates are laid side by side and are marked on their faces rather than edges.

Not-So-Rough Openings

BY JOHN SPIER

Although the name suggests otherwise, rough openings demand plenty of precision, especially when they are framed in load-bearing walls. Properly done, rough openings provide a place for windows and doors to fit securely, unaffected by the critical structural work being done by headers, king studs, trimmers, sills, and cripples (or sill jacks). After years of building, I've learned that getting rough openings right makes the rest of the job go smoothly.

Check the Plans First

Although the rough openings for doors and windows are specified on the plans, these dimensions are worth double-checking. It's important to note that sizes always are described width first, then height. Getting this information correct is the first step in avoiding serious frustration a few weeks down the line.

Occasionally, circumstances can require rough openings to be modified. Nonstandard floor thicknesses, specialized flashing elements, and applied sills are just a few details that can affect rough openings and should be thought through. If the building details are particularly unusual or complicated, it's smart to test the scenario as a mock-up before committing to a whole project.

Verify Rough-Opening Locations

In conventional platform framing, headers typically are sized so that the opening is at the correct height with the header tight to the top plate. Sometimes this placement needs to be modified, either by using cripples above the header or by moving the header up into the plates.

Lateral locations of rough openings usually are specified from the edges of the building to the centers of the openings, and between centers when several rough openings appear next to each other.

Before transferring layout marks to the lumber, I give the entire plan a final check. Confirm clearances, and make sure that the rough-opening layout will maintain symmetry within and between floors, if that's a priority. When there is room to move left or right, it's nice to make sure that trim details fit cleanly without ripping and squeezing.

Gather All the Pieces

All the wall components that define a rough opening need to do their part in making the opening solid and square forever. The king studs on each side of the header should be straight in all directions (for example, no bow, crown, or twist). The header needs to be sized appropriately and should provide room for insulation, if possible.

FRAMING TERMS

ROUGH OPENING
An opening deliberately oversize by ½ in. or so for windows and doors to be shimmed plumb and level.

HEADER
In load-bearing walls, this beam carries the load around windows and doors.

KING STUD
Full-length studs nailed to each side of the header to support the rough-opening assembly between plates. They are the same length as the wall's common studs.

TRIMMER STUD (ALSO CALLED JACK OR JACK STUD)
These studs are nailed to the king studs directly beneath the header and carry the load transferred by the header down to the bottom plate. Trimmers set the height of the rough opening.

SILL (ALSO CALLED BENCH OR SADDLE)
This 2×6 is laid flat and nailed between the trimmer studs to support the window. The sill serves as a foundation for the window's pan flashing. Adding a second sill here provides extra nailing for trim, but it is optional.

CRIPPLE (ALSO CALLED SILL JACK)
These shortened studs support the sill and act as nailing for sheathing and interior-wall finishes. They follow the same layout points as common studs for the entire wall. Nailing sill jacks to trimmers when they don't land on layout is optional.

Omit if rough opening is for a door.

GET EVERYTHING READY

ONCE THE LOCATION AND SIZE OF ROUGH openings are verified, I cut all the pieces before assembling the walls. This step minimizes work time and waste, and ensures a speedy and accurate process when it's time to assemble the walls.

ORGANIZE YOUR MATERIALS. The headers I'm using here are easy to assemble from conventional lumber. My wife, Kerri, aligns the pieces as I follow with the nail gun. We neatly stack and organize the sills, sill jacks, trimmers, and headers prior to wall framing.

Trimmers need to be continuous from the header to the bottom plate. In some areas of the country this detail is required by code. Even if it isn't required, I still maintain that this practice is the best. Interrupting the trimmers with sills is not a good idea, because the ends of the sills eventually can crush under the load. Multiple gaps, even of the slightest dimensions, can allow settling to occur. Lateral resistance of the wall is better with continuous trimmers as well.

The building code doesn't specify the number of trimmers, but convention (and my building inspector) requires that we have two at each end of openings over 6 ft., three over 12 ft., and stamped engineering on anything questionable.

The sills, which sometimes are called benches, need to remain flat and straight to support pan flashings under windows, and sometimes the jambs of the windows themselves. Cripples (sill jacks) support the sills and maintain the stud layout of the wall for fastening sheathing and wall finishes. Some framers use a single sill and skip the cripples against the trimmers when they don't fall on layout; eliminating three pieces of wood here saves time and materials and still meets code. I prefer the extra nailing and solidity of double sills with end support.

Assemble the Parts in Order

While assembling the parts, it's important to keep everything flush to the inside. I keep framing tight to the deck as I nail the pieces together, and I nail carefully so that split ends don't create bumps. Maintaining a flush inside surface helps to minimize drywall cracking and also creates a smoother final wall finish.

The rough opening goes together in clearly defined steps, and following this process in order is

FIXING WRONG ROUGH OPENINGS

Problem: The window opening is too short.
Solution: Enlarge the opening by removing one of the two sills. If there's only one sill, cut the tops off the sill jacks to lower it.

Problem: The header is too low.
Solution: Substituting a smaller but stronger header works, but then the trimmers need to be replaced.

Problem: The opening is too narrow.
Solution: If I need to gain ¼ in. or less, I use a circular saw to shave the trimmer studs. Anything the sawblade can't get I remove with a chisel. Substituting narrower trimmers is also a good option. After checking bearing requirements, I substitute 2× trimmers with 1× or 5/4 stock. Eliminating trimmers and substituting mechanical fasteners such as Simpson HH4 header hangers (www.strongtie.com) is an excellent fix. In the case of a window opening, replace at least the top sill with a longer one.

Problem: The opening is way too narrow.
Solution: Leave most of the framing in place, and simply move the king stud and trimmers on one side, replacing the header and sills with longer ones.

Problem: The rough opening is in the wrong location.
Solution: Rather than start from scratch, I move the rough-opening assembly as a unit. I start by cutting through the nails along both top and bottom plates. If the sheathing is on already, I cut the nails with a reciprocating saw after creating some space between it and the framing with a sledgehammer. Before moving the unit, I cut the new opening in the sheathing, then install the rough-opening assembly.

CENTERLINES GUIDE THE LAYOUT PROCESS

Most plans indicate door and window locations by noting a dimension to the centerline, so I start my layout by marking the centerlines on the wall plates. **1** Then I measure half the width of the rough opening on both sides of the centerline. **2** Using a Speed Square, I finish marking the trimmer and king-stud locations.

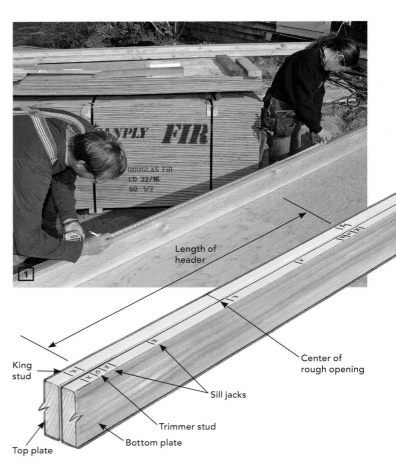

Length of header

Center of rough opening

Sill jacks

Trimmer stud

Bottom plate

King stud

Top plate

1½-in. mark on square

SQUARE THE WALL, THEN FRAME THE OPENINGS

Trimmer stud nailing pattern

I NAIL IN THE KING STUDS AND common studs first, leaving out any commons that will get in my way later when I'm nailing in the header. **1** Then I square up the wall and tack it to the deck to keep it that way while I finish the rough openings. **2** I install the header between the king studs next, **3** then nail in any commons that were left out. **4** Because doors and windows are attached to the trimmers, I nail them solidly to the king studs to minimize twisting and warping. I put two 16d nails at the top and the bottom, then one staggered every 12 in. in between. If it's a window opening, I don't nail between 10 in. and 20 in. off the floor. That way, plumbers and electricians are less likely to hit a nail when drilling. I also angle the nails slightly so that the points don't come through and damage hands and wires later.

important. Before I get into the main components of the rough opening, I nail the common and king studs through the top and the bottom plates starting from one end of the wall. For the time being, I leave out any studs or partition posts that are close enough to the king studs to prevent end-nailing the header.

Once the common and king studs are in place, I pin the bottom plate to a snapped line on the deck and square the wall. As a unit, the wall needs to be straight and square before the openings are assembled; once all the components are nailed off, any adjustment racks and bends the parts and loosens the joints.

TIP

I like to look up rough-opening dimensions in the tables provided by window and door manufacturers. These tables are excellent, and I've learned not to second-guess them.

CRIPPLES AND SILLS COMPLETE A WINDOW

WHEN INSTALLING THE CRIPPLES AND sills for a window opening, I nail the cripples into the bottom plate first. 1 Then I lay the sill across the cripples and transfer the layout. 2 I nail the first sill in place, then add a second sill (entirely optional) to act as nailing for interior trim. I keep the nails in the second sill aligned with the cripples to help drywallers and trim carpenters find the framing later.

1

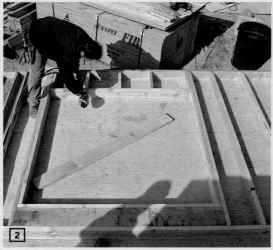

2

After the wall is squared, I drop the header into the opening and nail it securely through the top plate and then through the king studs. Before I forget, I like to nail in any studs that were left out to allow for clearance of the nail gun. The trimmers are installed next, using enough nails to keep them from twisting.

If I'm framing a window opening, the next step is to nail in the sill jacks. Put the end ones in first so that they can be nailed securely to the trimmers. To get an accurate layout for the first sill, I lay it along the bottom plate, transfer the layout marks, and nail in the sill. To give drywallers and trim carpenters a little help, I keep nails in the second sill in line with the sill jacks.

Complete the Final Step

Sheathing and housewrap come next, and I usually install these materials with the framing flat on the subfloor or slab. After this, there's one final step in finishing off the rough openings in the wall.

With the wall tilted up and braced plumb, I drive a few more nails in the corners that I wasn't able to reach while the framing was flat.

FINAL DETAIL. Once the wall is raised and braced, I finish nailing off the parts of the header that were not accessible while the wall was lying on the deck.

Framing Curved Walls

BY RYAN HAWKS

"Piece of cake," I replied the first time someone asked whether I could build curved walls. But as I looked over the plans, I didn't have a clue how to turn that drawing of a round room into reality. Late in the day, after the generators were silenced and the crews were on their way home or to the bar, I prowled around at a couple of jobs being built by one of the best framing contractors in San Diego.

By looking at his work, copying his notes from the floor, asking questions the next day, and applying ingenuity, I built my first radius wall as if I'd done it a hundred times. That was a few years ago; I've since developed a more refined technique that I use now.

A Chalkline Represents the Base of the Radius

The job shown here is a 180-degree radius wall, or a half-circle, that extends from two parallel sidewalls. This radius wall is the simplest type (see the sidebar on p. 175).

To lay out the radius, you first must establish the baseline. In this case, the baseline connects the ends of the two sidewalls. I snap a chalkline on the floor to represent the baseline, then mark its center (half the length of the baseline is the outer radius).

A nail partially driven into the floor at this center mark becomes the pivot point for my tape measure, which I use as a compass to mark the inside radius of the wall plate on the floor. Most tapes have a slot cut in the hook at the end for just this purpose: The edge of the nail head rides in the slot as you hold a pencil to the edge of the tape at the proper point and mark the radius to the floor.

Complications arise when something straight, such as a door or a window, goes into a curved wall. On this job, the architect had called for the wall to be 5½ in. thick and for the door to be 3 ft. wide. If I'd built it as drawn, the center of the door-jamb head would have intruded past the wall surface. After consulting the architect, we thickened the wall to 7½ in. and downsized the door to 2 ft. 8 in.

Thick Plywood Makes Up the Radius-Wall Plates

I usually make radius plates from two top and two bottom layers of 1⅛-in. plywood, which adds up to the same 4½-in. thickness as three 2× plates, avoiding the need to custom-cut studs. When framing on a slab, however, as here, the bottom plate must be treated plywood, which is available in a maximum thickness of only ¾ in. Cutting six ¾-in. plates to

(Continued on p. 176)

173

TO FRAME CURVED WALLS, use a tape measure as a compass and thick plywood for the wall plates.

START THE LAYOUT. Layout begins with a baseline and a centered nail. A chalkline between straight walls from which the curved wall will spring forms the baseline. A nail centered in this line is the centerpoint of the curve.

TAPE AS A COMPASS. The slot on the hook of the tape fits around the head of the nail, allowing the tape to be used as a compass, in this case to lay out the inside radius of the wall.

RADIUS-WALL VARIATIONS

LARGE-RADIUS WALL PLATES

When the radius is larger than 4 ft., the centerpoint can't be on the plywood. To find the centerpoint, snap a chalkline that's at least the length of the radius on the floor. Next, bisect the plywood with a chalkline, then align the plywood's chalkline with that on the floor. Locate the centerpoint of the radius along the chalkline, and use a tape measure to swing the radius.

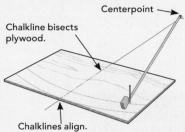

Centerpoint

Chalkline bisects plywood.

Chalklines align.

WHEN THE ARC IS LESS THAN 180 DEGREES

Jack radius walls, where the arc of the wall is less than 180 degrees, are common, but finding the radius of these walls so that you can lay them out is not so obvious. Jack radius walls are usually specified with two factors: rise and run. As an example, say the run is 8 and the rise 2. Plug these two numbers into a simple algebraic equation to get the radius (see the illustration below).

Radius = (Run² + 4Rise²) ÷ 8Rise
Radius = (8² + 4 × 2²) ÷ (8 × 2)
Radius = 5

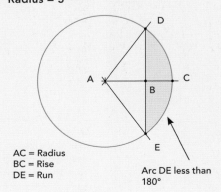

AC = Radius
BC = Rise
DE = Run

Arc DE less than 180°

FITTING DOORS AND WINDOWS

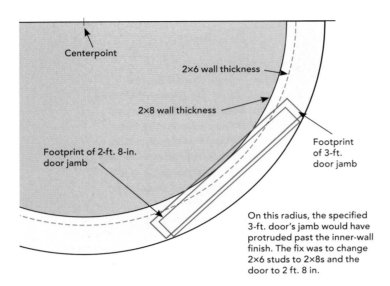

Centerpoint

2×6 wall thickness

2×8 wall thickness

Footprint of 2-ft. 8-in. door jamb

Footprint of 3-ft. door jamb

On this radius, the specified 3-ft. door's jamb would have protruded past the inner-wall finish. The fix was to change 2×6 studs to 2×8s and the door to 2 ft. 8 in.

FIT THE BOTTOM PLATE. The plates are marked to be cut using a tape as a compass.

MARK FOR DRILLING. Using a square to ensure that the plate aligns with the layout line, the author taps the plate over the anchor bolts to mark the hole locations.

achieve a 4½-in. plate thickness would have been a lot of work. Instead, I used two ¾-in. bottom plates and two 1⅛-in. top plates; I custom-cut the few studs that this wall required.

This particular radius was smaller than 48 in., so marking the plate radii on the plywood was straightforward. I simply set a nail in a line bisecting the sheet and used my tape as a compass to lay out the inside and outside radii on the plywood. (See the sidebar on p. 175 for walls whose radius is greater than 4 ft.)

If a radius is greater than 10 ft., I can cut the plywood plates with a 7¼-in. circular saw, but I cut these tight-radius plates more conventionally using a jigsaw. After cutting the first plate, I used it as a template to mark the other plates as well as the rough sills and the nailers at the top of window and door openings.

Once the plates are cut, I place the treated bottom plate on top of the anchor bolts, lining up one end with the intersecting straight wall. I use my square to align the plate with the radius I marked on the concrete and then hammer the plate over the anchor bolts to leave indentations. These indentations are where I drill for the bolts.

With the bottom plate drilled, I lay it back down on the anchor bolts to check the fit and to mark the ends to be cut where it meets the next section of plate. One end meets the straight wall, and its cut-line matches the baseline of the radius. If the other end is in the middle of the wall, I want this cut to fall in the center of a stud. I find this point by wrapping my tape around the outside of the plate and then marking the center of the stud that falls closest to the end of the plywood.

With the first plate section cut and in place, I repeat my actions to cut and drill remaining sections. Once the plate is in place, I detail the door and window openings on the plates.

The first step in laying out the window openings is to find their centers by measuring along the outside of the plate. With the rough-opening centers marked, I snap chalklines between them and the

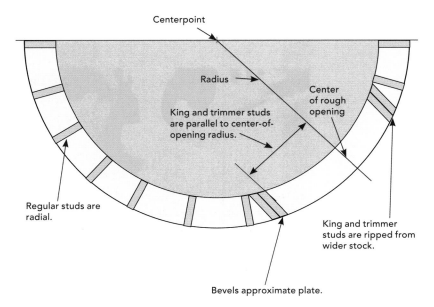

MARK THE NEXT SECTION. With the first section of plate drilled and placed, a square again is used to align a second section, and the author marks it for length.

LAYING OUT DOORS AND WINDOWS

The layout of the regular studs is radial, but the layout of the king and trimmer studs isn't. They must be parallel to a radius that goes to the center of the rough opening; otherwise the opening would taper and the window or door wouldn't fit. Kings and trimmers must be cut from wider stock.

Centerpoint

Radius

King and trimmer studs are parallel to center-of-opening radius.

Center of rough opening

Regular studs are radial.

King and trimmer studs are ripped from wider stock.

Bevels approximate plate.

MARK THE ROUGH OPENING. The author snaps a radius line to the center of the rough opening, then transfers that line to the bottom plate.

MARK THE TOP PLATE. The centerline is transferred from the bottom plate to the top (top), where the locations of the kings and trimmers are marked (above).

GETTING THE KINGS AND TRIMMERS RIGHT. The simplest way to find the width and bevel of kings and trimmers is to scribe a scrap block in place (above) and use it to set up a tablesaw to rip the kings and trimmers from wider stock (left).

centerpoint. Aligning my framing square on the chalkline, I use the square to mark the rough opening on the outside and the inside of the plate, then to connect the marks I've made. These layout lines aren't radial; rather, they're parallel to the center-of-opening radius. If they were radial, the rough opening would be narrower on the inside than on the outside.

Kings and Trimmers Are Custom-Cut

Because they are parallel to the centerline of their rough openings (as opposed to being radial), the king studs and trimmer studs are wider than the other studs, and their edges must be beveled to fit neatly on the plates. The bigger the rough opening, the bigger the bevel angle.

I find the rip angle and depth of the king and trimmer studs by laying a block on the plate where the king or trimmer stud sits and scribing. On this job, with its 2×8 studs, I was able to rip the kings and trimmers from 2×10 stock.

After marking the rough openings, I lay out the stud locations on the outside of the plate. Most architects note the stud spacing on the plans. Usually,

LAYOUT FOR STUDS. Studs on curved walls are often spaced closer than 16 in. on center, in this case 8 in. Laying out the outside edge of the plate ensures that the sheathing joints, though not the drywall, land on studs.

LINE UP THE STUDS. Snapping radial chalklines for each stud would ensure perfect alignment, but lining up the studs by eye works fine and is a lot faster.

THE HEADERS ARE STANDARD. Only the nailers for the wall finish (extra plate stock) are curved. Assembling the headers with a square as a guide ensures their fit.

the tighter the radius, the closer the stud spacing to provide backing for the wall finishes. On this job, the studs were on 8-in. centers.

The headers are standard, straight lumber that sandwiches between two pieces of 1⅛-in. plywood, extra pieces I cut along with the plates. I size this plywood by laying it on the detailed plates and marking it to fit between the king studs.

Nail the Wall Together Where There's Room to Maneuver

I like to frame these walls on a nice flat floor, but I framed this one outside because the job was a re-model with no room inside for me to frame a curved wall. Working with curved components is cumbersome, and it helps to have one person holding the wood and another one nailing.

I begin by nailing the headers to the king studs and trimmers, and then to the plates. Next come the studs, and after nailing in more than half of them, I chock the wall with a couple of scrap blocks to prevent it from rolling. The rest of the wall is filled in one stud at a time by two carpenters, one who is holding the studs and another who is doing the nailing.

Round walls are usually heavier than straight walls, and they like to roll. I always take extra precautions when standing a round wall; having too many carpenters raise a wall is better than not

having enough. When possible, I raise radius walls from the position of having the ends of the plates, as opposed to their centers, on the floor. This approach prevents the wall from rolling as it's being raised.

Because of the existing house, that approach wasn't possible on this job. There's really no trick here: Just have more muscle than wall on hand. Because the plates differ in thickness from straight wall plates, they don't lap to tie the walls together. They're simply butted, plumbed, and tied together with metal straps wrapped around both intersecting plates.

Sheathing the wall is simple. I use ⅜-in. CDX plywood because it's flexible, and I sheathe as I would any other wall.

TIP

It's possible to mark the studs exactly radial on the plates by pulling a chalk-line from the centerpoint to every layout mark. I usually don't bother with this step, however, finding that I can visually check the stud alignment well enough.

STUDDING OUT. Curved walls want to roll, so filling in the studs takes two carpenters. The author starts with the kings and trimmers (above). Next come the regular studs (left) and the midwall blocking, which he scribes from the laid-out plates (below).

TIE THE WALLS TOGETHER. Lack of plate overlap means metal straps must tie curved wall to straight. And yes, it would be safer to perform this operation on a ladder.

Framing Big Gable Walls

BY LYNN HAYWARD

Framing quickly and efficiently is all about doing things for a reason rather than doing things just to see wood go up fast. Many builders frame the roof first, then frame the gable walls under it. Although they seem to be getting a lot done fast (because the roof is up), it's actually very dangerous to dangle off a ladder while maneuvering full sheets of plywood and a nail gun, trying to sheathe the gable wall that just went up so quickly.

In the long run, I find that building the whole end of the house in place on the floor is much safer and more efficient. If the gable end has any kind of unusual trim detailing, as was the case on the house featured here, building flat on the deck makes even more sense.

Don't Snap Lines for Every Stud

Even builders who frame gable walls on the deck often expend more time and effort than they need to. A common practice is to snap the entire gable-wall layout on the deck: individual studs, top and bottom plates, and rafter layout.

You don't need to work this way, however. The only chalkline I snap on the deck is the line representing the inside edge of the bottom plate. Because I use the rafters to define the top of the wall (see the top photo on p. 183), any additional lines are superfluous.

My single chalkline is snapped 5½ in. in from the outside of the framing (for 2×6 walls). Rather than hooking the subfloor and then measuring in, I use a straightedge, such as a framing square, to plane up from the framing. Following this procedure keeps the outside walls consistently straight from one floor to the next.

I cut the bottom plate to length and toenail it to the deck through the outer face of the plate. The nail acts as a hinge when my crew and I are lifting the wall. If you toenail through the inside face, the nail will pull out of the deck and let the wall slide bottom first as you raise it. I also use short lengths of metal strapping nailed to the floor and to the bottom plate to prevent the bottom of the wall from kicking out during lifting. After the wall is in place, we pull these hinge nails and the nails in the straps with a cat's paw, then cut off the metal straps.

Rafters Replace the Top Plates

We eliminate the top plates in gable walls in favor of notching the studs to accept the rafters. This method makes stronger and lighter walls, and it uses less wood. We nail the kneewall posts at a right angle to the ends of the bottom plate and tack them to the floor to hold them in place while we build the gable wall. Atop the kneewall posts, we place the straight-

BUILD GABLE WALLS ON THE FLOOR, lift them with specialized jacks, and eliminate extra footsteps wherever possible.

BUILD THE OUTSIDE FIRST

To SHEATHE A WALL BEFORE STANDING it, the wall should be straight and square. My process begins at the bottom plate and ends at the rafters, which define the top of the wall. Because the wall in the photo at right gets a large French door, I omitted the central section of the bottom plate rather than cutting it out later. When doing this, it's important to toenail the plate in its exact location (see the photo below) to keep the rough opening plumb and square.

1. Toenail the bottom plate to the subfloor.

Snap a chalkline 5½ in. from the floor framing. To locate the line, use a framing square to plane up from the rim joist instead of hooking the subfloor, which may not be cut straight. Many framers toenail the bottom plate from the inside so that the toenail is easier to remove after the wall is raised. The reason why it's easier to remove the nail is exactly the reason why I do it differently: The nail pulls out as the wall is stood. If you toenail from the outside, the nail stays in the deck, which helps to prevent the wall from slipping out as it's raised. I go a step farther and nail metal straps to the wall and floor before standing the wall. After the wall is raised, it's fairly easy to pull the nails and remove the exposed portion of the straps.

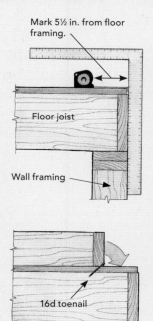

Mark 5½ in. from floor framing.

Floor joist

Wall framing

16d toenail

Metal straps

Toenail from the outside.

Toenail end of wall at rough-opening location.

2. Set the end posts of the wall.

If the gable intersects a kneewall, as pictured, the post height equals the kneewall height minus the bottom plate.

Continuous bottom plate if no rough opening

Kneewall posts

3. Set the rafters in place.

Use 4-in. spacer blocks to hold the rafters off the floor and flush to the face of the kneewall posts. Between the rafters' plumb cuts, set a 2×6 where the ridge board will sit.

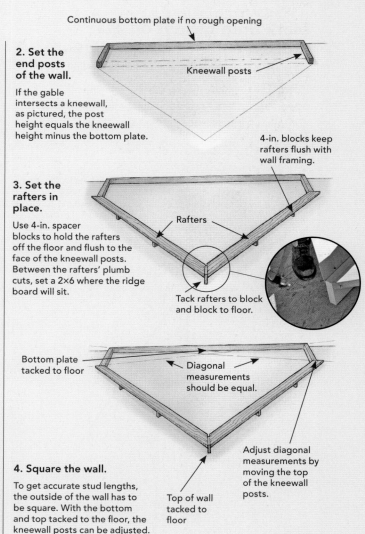

4-in. blocks keep rafters flush with wall framing.

Rafters

Tack rafters to block and block to floor.

4. Square the wall.

To get accurate stud lengths, the outside of the wall has to be square. With the bottom and top tacked to the floor, the kneewall posts can be adjusted.

Bottom plate tacked to floor

Diagonal measurements should be equal.

Top of wall tacked to floor

Adjust diagonal measurements by moving the top of the kneewall posts.

est two rafters in the pile and nail them down. At the peak, we tack a 2×6 block between the plumb cuts in the rafters with its bottom placed exactly where the bottom of the ridge board will be. A stud will be cut to fit below this block, and the block will be removed before roof framing. This puts the ridge at exactly the right height during roof framing.

With the outline of the wall nailed together, and the bottom plate and the ridge block tacked to the subfloor, we square the wall using the kneewall portion as our rectangle for equalizing the diagonals. When the wall is perfectly square, we tack the end posts in place to the floor and turn to framing the inside of the wall.

Mark Stud Layout on the Rafters with a Framing Square

After laying out the wall on the bottom plate, we transfer the wall layout to the face of each rafter using a framing square and a pair of screw-on stair buttons (see the sidebar on the facing page). In stair-

layout language, you have to position the buttons on the square so that the broader blade steps off a 16-in. run while the square's tongue remains plumb—or perpendicular to the bottom plate, because the wall is framed on the deck.

Before measuring the studs, we stretch a gauge string along the top of the rafter to see if there's a bow. If we find a bow, we cut the middle stud short by a little more than what the string shows. We then nail the stud on layout to the bottom plate, tack it to the floor with a toenail, and use a block (which is nailed to the subfloor) and a short length of 2×4 to pry the rafter straight. We face-nail the rafter to the stud and then release the lever. This step usually takes care of any discrepancies.

Cut All the Studs at Once, Not One Stud at a Time

To get the stud length, I hook my tape on a scrap of 2×6 tight against the bottom plate and measure to the long point of the beveled notch. It's better to

A STRONGER DESIGN AND A FASTER LAYOUT

Some framers build gable walls the same way they frame typical sidewalls: with a bottom plate, a double top plate, and studs in between. I don't do it this way. Instead, I skip the top plates and notch the studs to accept a rafter. This approach is stronger because it eliminates a hinge point by nailing through the face of the rafter into the stud. This method also uses less wood and is faster to assemble.

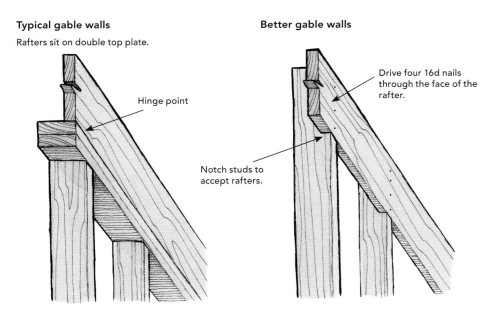

Typical gable walls
Rafters sit on double top plate.

Hinge point

Better gable walls

Drive four 16d nails through the face of the rafter.

Notch studs to accept rafters.

LAY OUT STUDS WITH A FRAMING SQUARE

WITH STAIR BUTTONS attached to the tongue and blade of a framing square, you can mark the gable wall's stud layout right on the face of each rafter. Button position depends on roof pitch. The 1½-in.-wide tongue of the square should remain perpendicular to the wall's bottom plate as you move up the rafter, stepping off a 16-in. run. Tracing against the edges of the tongue gives you the placement of the stud.

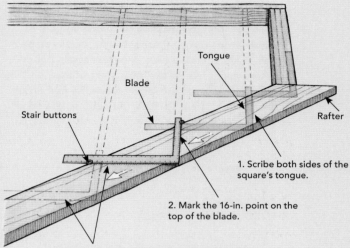

Tongue

Blade

Stair buttons

Rafter

1. Scribe both sides of the square's tongue.

2. Mark the 16-in. point on the top of the blade.

3. Slide the square up to the previously marked 16-in. point and repeat step 1.

measure to the long point, because after cutting the beveled notch you may need to hook your tape on the long point and pull the overall length.

As I measure the studs, I mark their lengths on the face of the rafter, then on a block of scrap wood. This way, I can cut all the studs necessary, or at least six or eight at a time. Because the notch in the stud is beveled, I need to mark the slope of the rafter on each stud, but I do this only once (with a Speed Square). For each subsequent cut, I use the offcut from the first stud as a scribing block.

You should be able to cut two gable-wall studs from each 16-ft. 2×6: a long and a short piece, or two medium pieces. The top of the stud is cut square below the top of the rafter. For each stud, I make two marks: One represents the square cut and the other is the long point of the bevel. I trace the 1½-in. width of the notch with the tongue of the framing square,

scribe, then cut the notch. As I cut the studs, I cross the lengths off the list.

Use a Router to Cut Out Windows, Doors, and Sheathing Edges

The framing is straightforward. Check the rafter against the gauge string and correct if necessary. Frame window and door openings in the center of the wall first, then work your way out. Finally, face-nail the rafter to the studs.

I begin sheathing by snapping a chalkline about 3 ft. up from the bottom plate. This overhanging sheathing covers the floor framing and ties the gable wall to the rest of the house. To get the exact overhang measurement, measure down from the subfloor to the top of the first floor's wall sheathing. Subtract ⅛ in. from this number. It's important

NOTCHING STUDS IS DONE QUICKLY

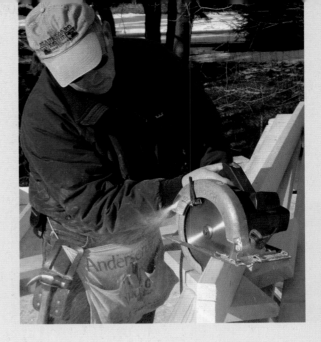

ONCE THE STUD LAYOUT IS MARKED ON THE rafters (see the sidebar on p. 185), the lengths of all gable studs can be measured and studs can be cut all at once to save time. Measure studs from the wall's bottom plate to the long point of the notch at the bottom edge of the rafter. Once the first notch is cut, use the offcut to mark the notch angle on all remaining studs.

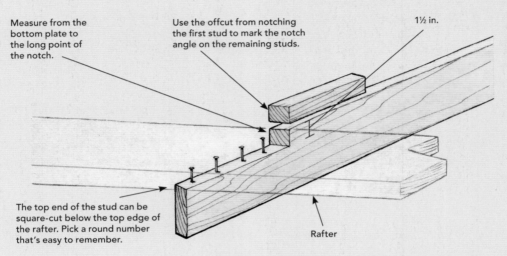

Measure from the bottom plate to the long point of the notch.

Use the offcut from notching the first stud to mark the notch angle on the remaining studs.

1½ in.

The top end of the stud can be square-cut below the top edge of the rafter. Pick a round number that's easy to remember.

Rafter

that the sheathing not hang down too far, or it will prevent the gable wall from resting on the subfloor, making it unstable.

As the framers sheathe the wall, I break out my 3-hp router with a panel-cutting pilot bit, and I cut out the window and door openings as well as the overhang along the top of the rafters. This router technique is much faster than marking, measuring, and snapping a bunch of lines to represent these cutouts and then using a circular saw to make the cuts. This technique is also a lot cleaner, leaving no sheathing bumps to hinder the window and door installation.

When installing the housewrap, we make sure to snap a line, similar to the one for sheathing, to allow overlap on the first floor. We also allow for about a foot of overlap at the corners, stapling along the corner and folding back the extra. As we progress along the wall, we pull the housewrap tight. It's important to keep the housewrap tight and to avoid big bubbles that can cause a problem when snapping lines for the clapboards or the siding.

I don't like to install windows or apply siding to a wall before standing it up because the racking that takes place during the lifting process can loosen the nails. I do, however, apply exterior trim to the gable.

INSTALL A DUMMY RIDGE BOARD. A short 2×6 provides the proper spacing for a ridge board to fit between rafters after the wall is raised.

SMOOTH FRAMING STRATEGY. Before you start filling in the wall with studs, run a string along the top of the rafter to check for straightness. If there's a substantial bow, bend the rafter back straight with a lever and a block nailed to the subfloor. Then secure it with a stud or two.

FRAME FROM THE INSIDE OUT. If you frame the window openings first, you'll be able to nail sideways into the header with a nail gun for a strong connection. Cripple studs above the header are placed last because they need to be measured after the window opening is framed.

I make sure the trim has been painted with at least two coats of exterior paint. Here's a tip about the gable fascia: As wood dries, it shrinks. Because the board shrinks equally from side to side, it shrinks the most along the short point of the miter, the widest part of the board. To minimize this problem, we cut the fascia so that it has a small gap at the long point. A couple of weeks later, after the short points have shrunk, the gap is uniform. This usually coincides with the end of the roof dry-in, and as long as we're up there, we toenail the gap together with a stainless-steel siding nail for a tight, long-lasting joint.

Wall Jacks Make Easy Work of Lifting Heavy Walls

After the wall is framed and sheathed with house-wrap and the fascia is applied, it weighs a lot. I use wall lifts made by Proctor®. They're basically steel pipes with a come-along mounted to them (they telescope for easy storage). At the bottom of the rod

WALL SHEATHING SPANS THE RIM JOIST. Hanging the sheathing below the bottom of the wall ties the two stories together structurally. Don't hang it down too far, though, or it will prevent the wall from sitting on the floor fully.

CUT OPENINGS QUICKLY WITH A PANEL-CUTTING BIT. I look for a plunge-cutting panel bit with a cutting length of 1⅛ in. or 1¼ in. This allows me to get twice the life out of the bit: After I dull the top half, I reset the bit lower for a fresh cutting surface.

is a hinged shoe; one leaf of the hinge is nailed to the subfloor and the other leaf is welded to the pole.

The pole can move with the wall as the wall goes up. The pole starts out standing basically plumb, and as the wall is jacked up, the top moves farther and farther away from the shoe, hence the hinge. At the top of the pole is a pulley through which the cable from the come-along runs. At the end of the cable is a bracket nailed into the top of the wall.

The wall jacks also have a small bracket at the top of the pole to stop the wall from falling past the 90-degree plumb position and falling off the house. After the wall is standing plumb, we attach braces to hold it that way, and we remove the wall jacks.

With the wall stood and braced, we pull the toe-nails and then cut the straps; then we nail the wall to the line. This sometimes takes a little persuasion.

A LITTLE PERSUASION GETS THE WALL BACK ON THE LINE. With the wall stood and braced, we nail down the bottom plate, adjusting it to the line as we go. We also drive a few 8d nails through the sheathing into the rim joist to squeeze it tight before nailing off the sheathing with a nail gun.

LIFT WITH YOUR HEAD, NOT WITH YOUR BACK

LIFTING EXTERIOR WALLS WITH WALL JACKS is safer in the short and long run: The wall doesn't fall on anyone during lifting, and my back lasts longer. The 16-ft. jacks consist of two telescoping steel poles, a come-along, and a hinged shoe. A bracket at the end of the come-along cable is nailed into the top of the wall, the hinged shoe is nailed into the subfloor, and the wall is raised by cranking the come-along. A hook on top of the poles keeps the wall from falling off the house after it has been stood.

Raising a
Gable Wall

BY JOHN SPIER

There is no easy way to set up the staging for building the gable wall of a house. Most approaches involve pump jacks, wall brackets, an assortment of ladders, and too much time. Even a simple gable can be tough to stage, and I've seen gables with their peaks 40 ft. off the ground and others that had to be built above another roof that were nearly impossible. One of the scariest staging jobs that I ever saw was for a gable that was built over a large greenhouse.

Fortunately, there is a way to avoid staging altogether: If you build the gable walls on the deck, you can raise them into place. Often, when I raise a gable wall, it has been framed, sheathed, sided, trimmed, and painted. The closest anyone has to get to that gable again is when the roofer looks over the edge.

The Gable Is Laid Out
on the Deck

In some cases the gable walls that I build and raise include a full-height wall or a kneewall below the triangular portion of the gable. For the Cape-style home shown here, the top floor was to be under a simple gable roof with dormers on both sides, so we built and raised only the triangular section of the gable wall.

After I've built the uppermost floor of the house, I snap chalklines for the perimeter walls. If the floor isn't perfectly square, I make a few minor adjustments to get the sides parallel and the ends square, a procedure that will simplify the roof framing immeasurably. The next step is to snap lines representing the peaks of both gable-end walls.

I start by snapping the centerline, or ridgeline of the house, and the tops of the sidewalls, if any. Then, using the distance from the centerline to the sides, along with the roof pitch, I calculate the height of the peak and snap lines for the top of the gable wall. These lines represent the bottoms of the rafters, which are also the top-plate lines. Even if the two gable-end layouts overlap in the middle of the deck, I snap them. As a final check, I measure the four top-plate lines. If they're not all the same, I figure out why and fix the problem.

At this point, I use the length of the top plate to lay out and cut a pattern rafter. I check it for fit by laying it down in place on the gable-end chalklines. I then make four rafters for the gable walls, label the pattern clearly, and set it aside for framing the rest of the roof.

FINDING THE PEAK. The first step is laying out the upper plates and snapping chalklines for other framing details.

FRAMING THE ENTIRE GABLE while it's flat on the deck is fast and easy. Raising it is a challenge and a danger.

BLOCKS KEEP THE PLATES STRAIGHT. Top plates are tacked to 2× blocks that have been ripped to 2 in. The blocks will also support the gable rafters.

METAL STRAPS ACT AS WALL HINGES. Nailed to the floor framing and then to the bottom plates, straps keep the wall from slipping off the deck's edge.

Steel Bands Act as Hinges

Before I start assembling the wall, I lay out a few more things on the floor. I locate any interior walls that intersect with the gable so that I can include partition posts. Any other items that require a nailer, such as a tub or shower or closet shelves, are also located and marked. Finally, I locate the windows and snap lines on the floor at the insides of the king studs. Now I'm ready to start framing.

First, I cut and assemble the plates. To hold the plates straight, I use short 2× blocks ripped to 2 in. for 2×4 plates and 4 in. for 2×6 plates and nail them to the deck every 6 ft. or so to hold the plates straight. The blocks are nailed to the floor on edge above the plates, where they serve double duty by temporarily supporting the gable rafters. The plates are then tacked to the blocks.

For the bottom plates, I use 12-in. ridge ties or short lengths of metal banding that act as hinges when the wall is lifted. The bands are nailed through the floor sheathing and into a joist, block, or other framing below. The plates are set on top of the steel band, and

BRING ON THE SHEATHING. The completed gable frame—including rafters, studs, and blocking—lies on the deck, ready for sheathing. Horizontal edge-blocking for the sheathing is required for this high-wind area.

the band is bent up and nailed to what will be the underside of the plate. These "hinges" are essential to stop the bottom of the wall from kicking out during the raising. After the wall is in position, I cut off the inner part of the band with a reciprocating saw.

Next, I install the king studs for the windows and add the trimmers, headers, sills, sill jacks, and cripples. If there is no window in the center, I install a full-length center stud or post to carry the ridge. If there is a center window, I put in the ridge post above the header. The studs are then laid out and installed for the gable wall using the same spacing that I established on the first floor, which keeps the framing neatly in line from foundation to peak. After the studs are cut and nailed in, I add blocking, fire-stops, nailers, and any other framing that might

need to go in the wall. If there are gable louvers or other vents, they are framed in, too.

The next step is adding the rafters that sit on the gable-end plates. I lay the rafters on top of the blocks holding the plates straight and nail the rafters to the plates. The top ends of the rafters get shaved slightly so that the ridge will slide in easily after the walls are up. I also make sure the bird's mouths will fit the sidewalls after the gables are raised, and I pull the nails holding the plates to the blocks.

Add the Sheathing, Housewrap, Trim, and Shingles

With the framing complete, I turn my attention to sheathing the wall. I leave the appropriate amount of sheathing extending beyond the bottom plate to

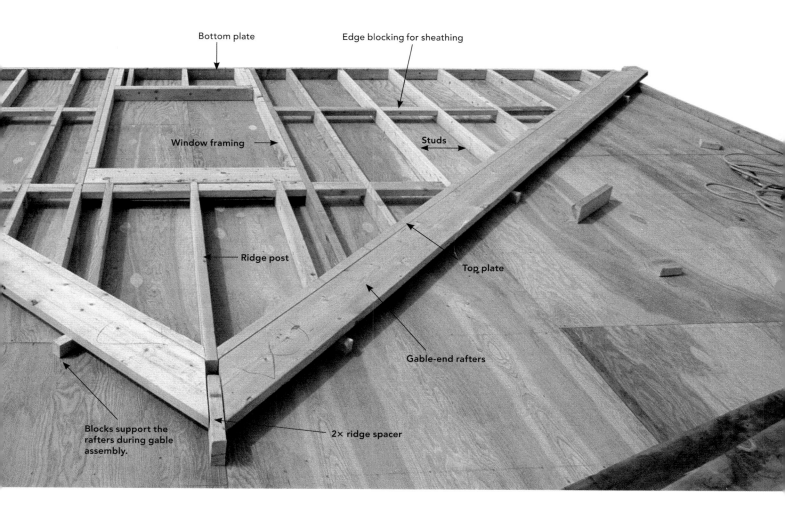

Bottom plate

Edge blocking for sheathing

Window framing

Studs

Ridge post

Top plate

Gable-end rafters

Blocks support the rafters during gable assembly.

2× ridge spacer

meet the sheathing of the wall below, subtracting ½ in. from my exact measurement to allow for shrinkage, settling, compression of the floor, and carpenter error. If there is a lot of loose plywood hanging out beyond the edge of the house, I mark it so that no one walks out and falls through.

I finish sheathing the entire gable and cut out window openings. As you might expect, nailing off the sheathing while the wall is lying flat gives us a big advantage in terms of time, effort, and accuracy. The housewrap also goes on much quicker and easier while the wall is horizontal.

After the housewrap, I nail on the felt-paper splines for the windows. Another strip of felt paper goes along the top of the wall, and sometimes a piece is needed for a box return or other trim detail.

Now I'm ready to tackle the rake detail. If there is an overhanging rake, I use the pattern rafter to cut the fly rafters, adding half the ridge thickness to each so that they meet in the middle. The fly rafters are installed on lookout blocks, usually at 24 in. on center. A framing square on the floor helps us to keep the overhang in the proper roof plane during assembly. There was no gable overhang with this

IT SURE BEATS CARRYING PLYWOOD UP AN EXTENSION LADDER. Sheathing the gable wall while it's flat means that sheathing doesn't have to be held in place while it's being nailed. Cutting and nailing also go much quicker horizontally.

GIFT-WRAPPED GABLE. With the wall lying flat, the housewrap can be stretched out and stapled without wrinkles.

FITTING GABLE TRIM IN COMFORT. Instead of working off tall ladders or awkward staging to get all the trim cuts right, the author fits all the trim perfectly with the wall lying flat.

GABLE VENTS FOR HIGH-WIND AREAS

THE STORMS THAT BUFFET BLOCK ISLAND are typically accompanied by gale-force winds blowing in off the water. Precipitation in these storms travels horizontally, forcing its way into even the smallest openings. Traditional louvered gable vents inevitably leak.

The bow vent (see the photo on p. 197) is a decorative but functional solution. This vertical eyebrow arrangement admits air for ventilation but shields the opening from rain or snow. Local codes may vary in determining vent size, but for this job, we cut 2-in. by 24-in. openings on both gables. We then staple copper or bronze screen over the opening. Two bow-shaped nailers are then cut out of pressure-treated or cedar 2× stock—one large bow about 2 ft. wider than the opening and another proportionately smaller bow, both with slight angles on the tops for the angle of the shingles. The center of the larger nailer is cut out, leaving a hole the same size as the vent opening. The solid upper nailer stiffens the shingles and limits the space in the vent to discourage birds and insects. The nailers are then attached to the gable wall above and below the opening. After shingling up to the lower 2× nailer, we shingle over both nailers starting in the middle and working to the sides. A starter course under the bottom course ensures a weathertight vent. We continue shingling to the peak, blending bow shingles with wall shingles.

These vents work best in conjunction with a small attic space, soffit vents, and vent baffles. When correctly sized and built, bow vents provide adequate ventilation even on fair-weather days.

A HOLE FOR AIR BUT NOT WATER. A bow vent begins with a hole cut in the wall properly sized for the area to be vented.

THE BOW IS FRAMED WITH 2×S. Beveled nailers are attached above and below the vent opening.

SHINGLES FOLLOW THE LINE OF THE BOW. Shingles are nailed to the nailers using a stretched string as a guide.

project, so we simply nailed a subrake on top of the felt-paper strips.

At this point, we can install most of the trim. If there are cornice returns, we build them and start our trim there. If not, we just leave the rake boards long to be cut and fitted when the rest of the roof is framed and trimmed.

To save weight for the lift, we usually leave out the windows. A course line for the shingles is established above the finished window height, and we shingle up from there. (For smaller gables, we install the windows and shingle the entire wall.) We nail the lowest course near the tops of the shingles so that the courses below can be slipped in underneath. (An inconspicuous row of stainless-steel nails will secure them later.) We also build our bow vents at this time. Finally, if time and weather permit, we fill nail holes, caulk, and put three coats of paint on all the trim.

The Gables Are Raised in Stages

Some of the heaviest gables we've ever stood up were 30 ft. long and 16 ft. high, and framed with 2×6s. Usually about six or eight people are required to raise walls such as these. Of course, a crane or commercially available wall jacks could make the job easier, but I don't happen to have either. The trick is not to raise a really big or heavy wall all in one motion.

We start with a couple of strong sawhorses standing by and lift the peak enough to kick the horses

BRACES HELP WITH THE LIFT. With the wall held in place at about a 45-degree angle, 2× braces are nailed high up on the studs (above). The braces then help to control the final push (right) and finally are nailed to blocks attached to the floor framing.

underneath. For this stage, everyone is lifting close to the peak for maximum leverage. With the wall resting safely on the horses, we next get a couple of 2×6 studs ready for props and lift the wall enough to prop it at the window headers, usually about 45 degrees. The gables shown here were small enough to be lifted directly to the 45-degree position.

At this point, the heaviest lifting is done, and control becomes our biggest concern. We nail in two or three braces near the top of the wall. The braces have to be long enough so that they will be at a 45-degree angle with the wall standing plumb. By angling several spikes through the brace and into a stud, a strong attachment is made that still allows the brace to pivot as the wall is raised.

TYING THE GABLE TO THE HOUSE. Once the gable is braced in position, the bottom plate is nailed off (right) and the sheathing that was left overhanging is nailed to the house framing (below).

The final lift is made using the braces to help push and control the top of the wall. Once it is upright and has been plumbed, the braces are nailed off to blocks nailed through the floor sheathing and into the joists. Sometimes we add a few additional braces to keep the rakes straight or for peace of mind overnight. When the wall is up and braced, the bottom plate and the overlapping part of the sheathing can be nailed off, tying the gable to the rest of the house.

A Couple of Additional Tips

If the peaks of the gable walls overlap by a little in the middle of the floor, we often build the first gable and lift it onto the sawhorses so that it is out of the way for building the second. This method allows both gables to be painted at once and means assembling extra hands for just one big lifting party.

Keep an eye on the wind while you're lifting the gables. If you're lifting into the wind, it's no problem. Just don't get squashed if you lose control of the wall. If the wind is behind you pushing the wall up and out, secure a safety rope to stop the wall before it goes beyond plumb.

Last, when the siding and painting subs show up, drive a hard bargain. The toughest part of their jobs has already been done.

Better Framing with Factory-Built Walls

BY FERNANDO PAGÉS RUIZ

Before I started building houses with factory-made walls, my five-man crew needed two weeks to frame a house. Today, the job gets done in just five days, with a crew of three.

Building with factory-framed walls demands good planning and an organized approach to the work that is done on site, which I'll discuss ahead. Once you make these adjustments, you'll be surprised by the benefits you discover (see the sidebar on p. 200). It wasn't easy persuading my crew to switch to factory-built walls, but now they wouldn't think of building any other way.

Factory-built wall panels can be ordered in different ways. You can get them framed with or without sheathing; you can have the siding installed; you can even have wall panels delivered with the windows installed. But windows break, and finished surfaces like siding can be damaged easily. So I build all my

DELIVERED TO YOUR SITE, prefabricated panels can save you time and improve quality on production houses or a custom house.

Seven Reasons to Let Someone Else Frame the Walls

1. A SMALLER CREW
One skilled framer and two helpers can erect factory-built walls faster than a five-member crew could build them.

2. A FASTER JOB
An average house constructed with factory-built walls can be fully framed in five or fewer days.

3. THE QUALITY IS UNBEATABLE
Every stud is cut with radial-arm-saw accuracy. In the factory, it's easy for workers to frame rough openings precisely and to nail every square inch of sheathing without missing a stud.

4. WEATHER IS NOT A CONCERN
Rainy days don't slow production, and the controlled environment of the factory means framing lumber and sheathing stay clean and dry throughout the process.

5. PRICES ARE STABLE
Because the factory builds so many houses, the manufacturer can order lumber and pass along some of the savings to builders.

6. MINIMAL WASTE
Offcuts are minimal; so is the amount of unusable lumber. As a result, the site stays clean, and your waste-disposal expenses are trimmed significantly.

7. SPECIAL DETAILS ARE NO PROBLEM
Before the factory builds the walls, you can edit the computer-generated plans. It's easy to eliminate unnecessary studs, add blocking for towel racks or cabinets, or make other framing alterations.

houses using factory-framed and sheathed panels for exterior walls, and factory-framed (no sheathing) panels for interior walls. Wall panels can be fabricated to any length, but they typically come in 12-ft. to 16-ft. lengths that a small crew can handle easily.

Computers Engineer the Walls

Building with factory-made walls is a lot like building with roof trusses. Engineers at the plant plot your plans into a computer, and unless you specify otherwise, their software automatically calculates the engineering values for headers, posts, and shear walls. The resulting wall-by-wall drawings provide a distinct advantage over conventional framing plans because you can use them to edit the framing with unprecedented detail.

I build the same house plans over and over again. For me, the advantage of this system is that I can incorporate what I learn during the construction of one house into the plans for the next house. For example, if the plumber says the framing over a sink is blocking a vent, I can make a note and forward the change to the factory. But even on a custom home, you can look at the factory's framing plans before they build walls and specify detailed information like hold-down locations, complicated shear-wall nailing, or blocking for kitchen cabinets. The plant's computer then makes sure all these details are framed in the right places.

Plan Ahead to Get Walls on Time

The only drawback and the greatest learning curve with factory-framed walls is ordering the walls and timing their delivery. You can't just call the plant and get walls at a moment's notice. You have to requisition walls in advance to get them delivered on time.

Some building contractors wait until the foundation has been poured before ordering walls. The factory then can send a representative to measure the foundation and adjust the plan dimensions accordingly. If you can afford the downtime, this is the safest way to do the ordering.

I don't like to add idle time to my construction schedule, so I order walls before breaking ground. The factory builds according to my house plans without verifying any of the job-site measurements. This approach means I have to make sure the foundation is perfect. But the extra effort is worthwhile because the walls arrive on the job the day after the slab goes in. Of course, if the house has a basement or crawlspace foundation, I don't want the walls to arrive until the first-floor deck has been framed and sheathed.

If you give two weeks' notice, you should have walls on schedule. But keep in mind that your framer must be just as punctual. When the walls arrive, someone must be there to unload and organize them. It's important to make sure each wall section is placed in a convenient spot.

It's Up to the Framers to Set the Panels Square, Plumb, and Level

When the delivery truck arrives, framers have to unload the walls in sequence and stack them in order of use: interior walls in the center, exterior walls at the perimeter, and second-story walls on blocks out of the way. The placement of second-story walls is critical because there's little room to move around large components. Factory components come labeled, so all you have to do is refer to the factory's plans to find out which walls go where.

Just as in conventional framing, you start by snapping lines on the foundation or the floor deck. You then stand up perimeter walls, nail off double plates, and install temporary braces. Next, you erect the interior walls, taking care not to box a panel section into a room, kind of like not painting yourself into a corner.

The most critical step is lining up the top plates, then squaring and straightening the walls. Because the panels typically come in 12-ft. or 16-ft. sections, you have to mate several to create a long wall. To keep the panels level, the top plates must line up flush before you nail two walls together. Sometimes you need to shim the bottom plate of one wall to level it with the adjacent panel.

One of the nice things about factory walls is that the top plates come marked at every intersection. This makes light work of measuring and fitting the double plate while leaving cutouts for intersecting walls. Most of the double plates come preassembled, but you have to cut and fit plates over wall joints.

On exterior panel joints and corners, the sheathing overlays the adjacent panel to create a structural splice. It's important to remember to nail the sheathing on these corners and laps. My framers nail off every panel before moving to the next wall section.

Once all the perimeter walls are up, it's important to square the house and align the plates with a string. After the walls are plumb and true, we brace them with 2×4s and continue with the second floor.

Are Factory-Built Walls the Future of Framing?

Just as roof-truss manufacturers have established themselves by providing one-stop engineering and production services, wall manufacturers can deliver an integrated structural package. This proves most beneficial to builders in earthquake and hurricane regions, where pre-engineered shear-wall systems can speed up job-site construction and reduce the need for elaborate hold-down systems.

Hardware companies, wall manufacturers, and truss plants have teamed up to develop a whole-house systems approach to framing, which promises to coordinate the delivery of a pre-engineered component system suitable to any hazard area from Northern California to Southern Florida.

I advocate using advanced framing techniques that reduce lumber consumption. One obstacle to implementing such techniques is training framers on the subtleties of optimized structural framing and energy-efficient detailing. Because the panel factory provides custom-engineering services, I can design walls using these advanced techniques.

One day, factory-built walls may become as conventional as roof trusses. Until then, I thrive on the competitive advantage that wall panels provide me.

Setting Wall Panels

First, get adjoining panels level, flush, and plumb. Then brace each section securely.

SHIM THE BOTTOM PLATE. To correct imperfections in the floor framing, drive a wedge under the bottom plate. When the top plates are level, nail the wall sections together and remove the shim.

ADD THE DOUBLE PLATE OVER WALL INTERSECTIONS. Make sure to bridge the intersection of two wall panels with a double top plate. The top plates are marked for intersecting walls (right).

NAIL THE END STUDS TOGETHER to join wall sections. When the top plates in adjoining panels are level and flush, nail the end studs together with 16d nails.

NAIL OFF THE INTERSECTIONS FROM OUTSIDE.
The wall sheathing extends over intersections and corners to create a structural splice. Nail the sections together from the outside.

PLUMB AND BRACE EACH WALL SECTION As you set the wall sections, check them for plumb and temporarily brace them with a 2×4 every 8 ft. to 12 ft.

Three Common Mistakes and How to Avoid Them

REVERSE PLAN ORIENTATION

If you forget to tell the panel factory that you have decided to reverse the floor plan (garage on the left instead of on the right), you have just committed a common error that can't be fixed inexpensively. The walls will be framed backward and inside out, with the sheathing on the interior face of the exterior walls. Once you finish fixing this "little" oversight by flipping the walls, reframing the openings, and stripping, then reinstalling the exterior sheathing, you could have framed the house twice.

CHANGES AND OPTIONS

If you send the factory your plans and then negotiate changes with your customer, don't be surprised to discover that the factory already has built your walls by the time you submit the modifications. If the changes are minor, it's not difficult to remodel walls on site, although you defeat the advantages of using factory-built walls.

FOUNDATION FUDGES

Before the factory-built walls are delivered, take the time to check the foundation for square and level. It's easier to deal with dimension and squaring mistakes than the ills of poor leveling. You can fix slightly out-of-square foundations by building the floor deck slightly oversize in both directions. This allows you to square off the deck and lose the foundation error. And you can add plywood strips as spacers between wall sections to gain an inch or two.

A foundation that's not level presents a bigger problem because the walls come off the assembly line square and true. Exterior walls won't conform to dips in the slab or stemwalls. You'll have to fix these problems by shimming the mudsill.

Shear Walls

BY ROB YAGID

Not every house needs to have shear walls integrated into the framing, but many do. In earthquake country, for example, shear walls help to strengthen houses so that they're far less likely to move under the severe lateral forces of a seismic event. Shear walls not only help to prevent catastrophic collapse; they also help to prevent smaller-scale damage like cracked drywall and fractured tile. Shear walls play the same role in houses in high-wind zones. No matter the source of the force exerted on a house—atmospheric or tectonic—shear walls are simply designed to protect the home and its occupants.

A key component of seismic-retrofit work is the integration of site-built shear walls into the framing. To be able to construct a shear wall so that it performs properly and offers maximum strength, you need to know how it works.

Performance under Pressure

Shear walls are designed to resist several forces simultaneously (see the top drawing on the facing page), and those forces can shift in opposing directions at any given moment. Here's an example of what can happen when a conventional wall experiences the stress of an earthquake or hurricane.

Lateral. The primary lateral force from an earthquake or high-wind event causes simultaneous uplift, compression, and sliding forces.

Uplift. Lateral forces try to roll the wall off the foundation, creating uplift on one end of the wall assembly.

Compression. As one end of the wall is experiencing uplift, the opposite end is under compression. These loads alternate as the building shakes back and forth.

Sliding resistance. The few anchor bolts that are present try to counteract the lateral force, which tries to slide the wall off the foundation, but the bolts are ineffective.

One Way to Build a Shear Wall

Extra foundation hardware, 4×4 posts, structural plywood, and a lot of nails help walls to resist the forces of earthquakes and high winds (see the bottom drawing on the facing page). These components shouldn't be added to a wall without the advice of an engineer, however. An engineer will optimize a shear wall's design to meet the specific demands of a house, which will dictate details like nail size and nailing schedule, hardware placement, and blocking size and orientation.

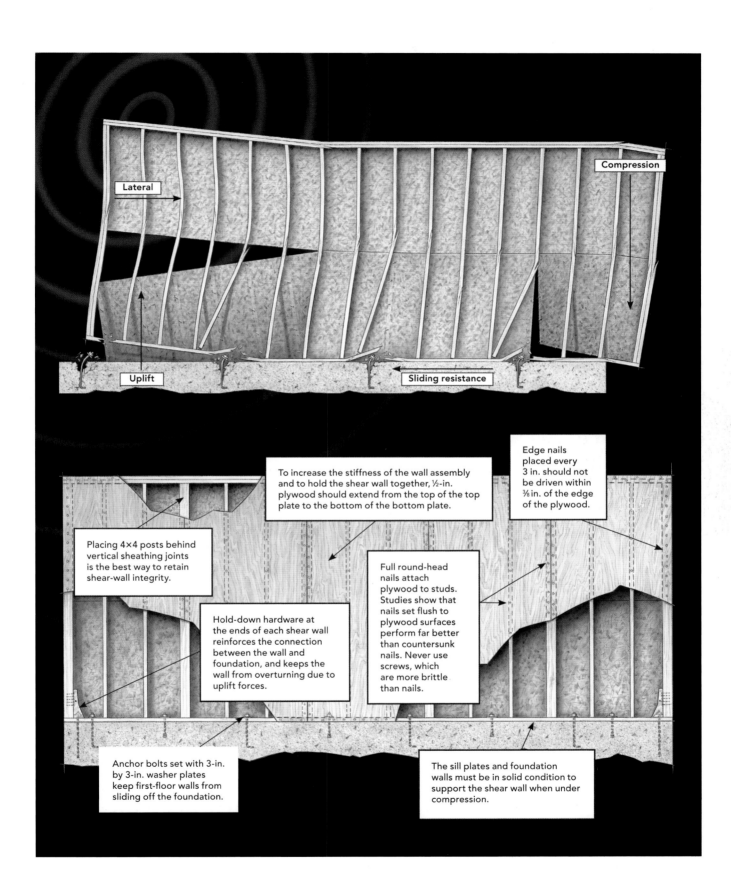

Lateral

Compression

Uplift

Sliding resistance

To increase the stiffness of the wall assembly and to hold the shear wall together, ½-in. plywood should extend from the top of the top plate to the bottom of the bottom plate.

Edge nails placed every 3 in. should not be driven within ⅜ in. of the edge of the plywood.

Placing 4×4 posts behind vertical sheathing joints is the best way to retain shear-wall integrity.

Full round-head nails attach plywood to studs. Studies show that nails set flush to plywood surfaces perform far better than countersunk nails. Never use screws, which are more brittle than nails.

Hold-down hardware at the ends of each shear wall reinforces the connection between the wall and foundation, and keeps the wall from overturning due to uplift forces.

Anchor bolts set with 3-in. by 3-in. washer plates keep first-floor walls from sliding off the foundation.

The sill plates and foundation walls must be in solid condition to support the shear wall when under compression.

A Slick Approach to Straightening Walls

BY ROE A. OSBORN

The framer's version of the classic chicken-or-egg question relates to straightening walls. Is straightening the final step in first-floor wall construction or the start of the second-floor deck? I think it's the latter, and here's why. Let's say you finish framing the walls on a Thursday. Friday it rains, so you don't work. You're off for the weekend, and Monday is a holiday. If you'd straightened the walls on Thursday, they would have had four days to move around in the wind and weather, and you'd probably have to tweak them again before framing the floor above. The floor framing locks in your straightening efforts, so that's why I associate straightening walls with framing the second floor.

The project shown here is a good illustration of how to straighten walls. Built to engineers' specs that satisfy the 110-mph wind-zone requirement, this framing is much beefier than normal, and therefore a little tougher to push back into a straight line. With this technique much of the work is a one-person job.

Create a Reference Line with String

Over the years I've seen lots of methods, special tools, and jigs for straightening walls, but in my opinion the springboard method I learned when I first started building houses still works best. It's a two-step process that starts with stringing the walls. Begin by nailing 2× blocks to the inside corners of all four walls. Then drive two additional nails partway into each block, as shown in the photo on the facing page. These nails act as anchor pegs for the string. Many framers use strong mason's twine for stringing because it can be stretched extremely taut. A chalkline can work just as well, though, and it has the advantage of having a hook on the end that can slip over the bottom nail on one block.

Tie or hook the string to the bottom nail and lead it over the top nail. Then stretch the string as tight as possible to the block at the other end of the wall. At this end, lead the string over the top nail, pull it tight, and wrap it several times around the bottom nail. Then wrap the string back over itself on the nail to keep it tight. The trick here is not to tie a knot that might have to be untied later. You now have a straight reference line running the length of the wall.

Let Kicker Boards Do the Work

Remove the temporary bracing used to hold the wall upright. (Unless conditions are very windy, even a fairly long wall should stand on its own for now.) The trick is to push the wall in or out to make it perfectly parallel to the string, which is where the springboards come in. For springboard material,

A ONE-MAN JOB. To straighten the wall, 12-ft.-long springboards are tacked underneath the top plate, then bowed down and tacked to the deck. Shorter kicker boards nailed beneath the springboards are used to manipulate the curve of the springboards, which in turn move the wall in or out. A taut string held off the top plate by blocks creates a guide that can be checked with 2× scrap.

Labels in image: Top plate, Stringline, 2× blocks, Gauge block, Springboard, Kicker

I use 12-ft. 1×8 rough-sawn pine boards because they're flexible, strong, and inexpensive. Also, they come in handy around the job site after they've fulfilled their springing duties. Taller walls require longer boards.

Walls are usually straightened one at a time, and it really doesn't matter which one is first. Choose a wall and position springboards every 8 ft. or so along the wall. Long headers at rough openings for windows and doors may require a springboard at each end. Also, be conscious of any hinge points, such as sheathing joints, that would make the wall bend. This is particularly important with tall walls.

Starting at one end of the wall, nail one end of a springboard to the underside of the top plate. Secure the other end to the deck, giving the board a slight downward bend as you nail it. This actually pushes on the wall, which means you'll likely be letting in the wall later.

Now nail the bottom of a 4-ft.-long 1×8 kicker to the deck below the springboard. Bring the top of the kicker snug against the springboard, but don't nail that end yet. Pushing the kicker board in or out changes the amount of arc in the springboard, which in turn moves the wall in or out.

Once the springboards are in place, you can begin to straighten the wall. Slide a 2× gauge block up to the string, then push on the kicker until the gauge block just slips under the string. Then drive nails through the springboard and into the end of the kicker to hold the wall straight.

After working down the length of one wall, sight the string and plate for a final check. This is your last opportunity to make sure the walls are dead straight before locking them in with the second-floor framing. The slightest deviation in the wall can turn into a major wave once the exterior siding and trim are applied.

START AT THE TOP PLATE. Placed at approximately 8-ft. intervals, springboards are first nailed to the underside of the top plate.

The top plate pushes out.

Push down on the springboard.

LOAD THE SPRINGBOARD. With the board's lower end on the deck, push down on the middle of the board, and tack that end in place.

INSTALL THE KICKER. The board that moves the springboard is known as a kicker. Typically a 4-ft. length of 1×8, it's positioned near the center of the springboard. Tack one end to the deck, and bring the other end up snug against the springboard.

The top plate pulls back in.

Push down on the kicker.

GAUGE THE MOVEMENT OF THE TOP PLATE. While sliding the kicker against the springboard with one hand, check the gap between the string and the top plate with a 2×4. Sliding the kicker away from the wall should pull the wall toward the line.

4

5

NAIL IT HOME. When the gauge indicates that the wall section is straight, drive a nail down through the springboard into the kicker to lock its position.

6

SOMETIMES YOU NEED MORE LEVERAGE

IN MOST CASES, SPRINGBOARDS CAN EASILY straighten a wall. But sometimes more force is required, especially near the end of a wall. A site-built lever lets you apply that force in a controlled fashion. Nail a diagonal 2× brace to a stud near the top of the wall at the trouble spot, and nail a long 2× block to the deck next to the loose end of the brace. Now nail a 2× lever to the block and to the brace. Pull back on the lever as someone else gauges the string. When the wall is straight, nail the bottom of the brace to the block to hold the wall in position.

DOUBLE-CHECK THE STRING. Once all springboards have been adjusted, go back and sight down the stringline to make sure that the wall is straight. Now is the time for any last-minute tweaks, before the top plates are locked into position by the joists above.

Framing Ceilings

Curved Ceiling?
No Problem

BY MICHAEL CHANDLER

A cathedral ceiling can open up a room dramatically, but if the ceiling is framed with a massive structural ridge beam, the beam will be a challenge to hide. An obvious solution is either to use bigger rafters or to fur down the ceiling to hide the ridge. Oversize rafters are a waste of wood, so my three-person crew opts for furring. As long as we're installing furring, why not have fun and curve the ceiling? The furring turns a chore into a delightful, economical upgrade. Adding 1×4 furring to the framing package costs less than increasing the rafters to 2×12s or I-joists. We can install the backing and the curved furring on a 27-ft. by 27-ft. ceiling in about three hours.

Strike a Curve, and Locate Backing

We use 1×4 #2 spruce furring. It's available in lengths up to only 16 ft., so most jobs require more than one piece to span the ceiling. To control the curve of the 1×4s and to support the ceiling, we fasten backing boards across the rafters, spacing them 3 ft. to 4 ft. apart. Just as when installing strongbacks, different combinations of dimensional lumber can be used for backing, depending on the offset required from the bottom edge of the rafter to back up the 1×4 curve.

On the project shown here, we bent a 16-ft. 1×4 between the end wall and the ridge beam and simply traced the resulting curve on the gable-end drywall. Measuring down from the rafter to the curve gave us the distance that the backing boards would need to span.

As shown on pp. 212–213, the resulting curve isn't a true arc, but a curve with flattened ends. We add fire-blocking in the walls where the curve dips below the double top plate.

Next, we slide short lengths of backing along the rafters to determine the size and location of the actual backing boards. Then we snap lines on the underside of the rafters to guide the installation. After assembling the backing boards on the floor, we attach them to the underside of the rafters with ring-shank nails or screws.

Offset the Splices in the Strapping

With the 1×4 furring strips, make sure to offset the splices in adjacent courses so that the overall curve of the ceiling can stay as fair as possible. Generally, the fairest curve should be across the center of the ceiling span, so it's smart to start with the clearest stock centered across every other rafter. This leaves short sections at the walls to fill later. Next, we

TRACE THE CURVE ON THE GABLE WALL, THEN INSTALL BACKING AND FURRING

Step 1: Butt, bend, and scribe

Butt a 1×4 up against the wall blocking, and then bend until it rests squarely against the ridge beam's bottom edge. Trace along the top of the 1×4, marking the curve on the gable-end wall.

Step 2: Install backing boards

Slide a short section of backing along the bottom of the rafter until it intersects the curved line. Attach the full-length backing boards here with ring-shank nails or screws.

Step 3: Nail up furring strips

Start at one end of the room by centering a 1×4 across the ridge beam. Nail the 1×4 to the ridge, then bend and nail it against the backing boards, using ring-shank nails. Butt furring joints over backing boards, and stagger joints in adjacent courses.

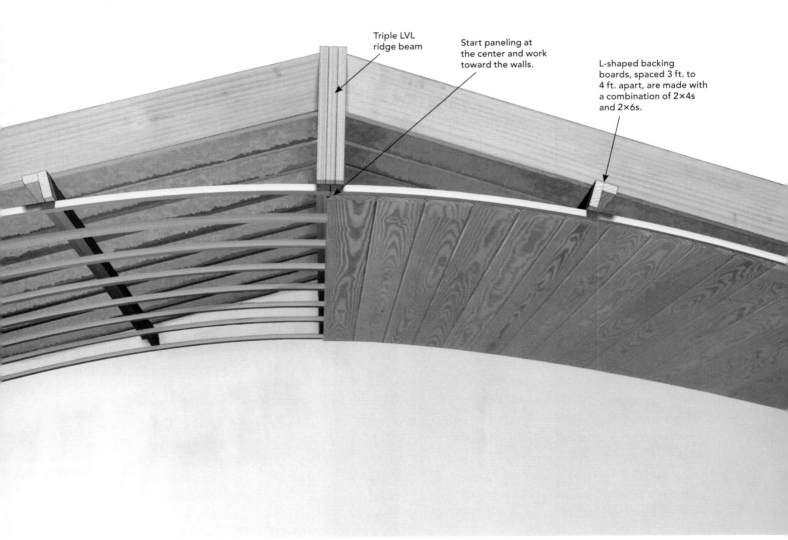

Ridge

Rafter

Configure backing boards to span between rafter and curve.

1×4 furring, spaced 16 in. or 24 in. apart

Triple LVL ridge beam

Start paneling at the center and work toward the walls.

L-shaped backing boards, spaced 3 ft. to 4 ft. apart, are made with a combination of 2×4s and 2×6s.

install long pieces starting at the walls, spliced on the ridge, between the first set of curves. We cut and fill the short end pieces with knottier stock.

Using ring-shank nails (or even deck screws) to fasten the furring strips is worthwhile because the nails will be loaded in withdrawal from fighting the tension of all those tortured 1×4s. We often finish the ceilings with wood paneling, but two layers of ⅜-in. drywall work, too. When you're using tongue-and-groove paneling it is important to control bulk air and humidity movement through the cracks between the boards as well as at the perimeter ceiling-to-top-plate transition. A conditioned attic with spray foam insulation solves this by reducing pressure and vapor differentials across the ceiling plane. With vented attics we have had success cover-ing the furring strips with drywall, taped and sealed to the adjacent wall plates, as well as with inexpensive 6 mil clear poly sheeting lapped down over soft foam gaskets on the adjacent top plates.

WOOD PANELING ADDS DETAIL. You can finish the ceiling with wood paneling or drywall.

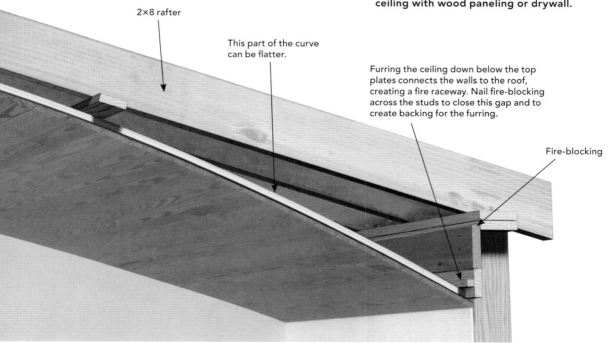

2×8 rafter

This part of the curve can be flatter.

Furring the ceiling down below the top plates connects the walls to the roof, creating a fire raceway. Nail fire-blocking across the studs to close this gap and to create backing for the furring.

Fire-blocking

Framing Cathedral Ceilings

BY BRIAN SALUK

I started framing houses years ago, before cathedral ceilings came into fashion. When asked to frame my first cathedral ceiling, I went at it much as I did any other roof. After bracing the walls plumb with leftover 10-ft. 2×4s, my crew set the ridge and the rafters, seemingly without a hitch. It was just another gable roof, only without the ceiling joists. When we finished setting the rafters, it was lunchtime.

I remember biting into my ham-and-cheese sandwich and looking back at the roof. I expected that feeling of satisfaction one gets looking on the results of a good morning's labor. Instead, I got a sinking feeling: The center of the ridge was sagging.

Back in the house, I saw that the weight of the roof pushing out on the walls had actually pulled some of the braces from the floor. Fortunately, it took only a couple of hours to jack the ridge level and pull the walls straight with a come-along. But I was lucky that the only serious loss that day was my uneaten lunch.

Brace the Walls to Resist Roof Thrust

The strength of a typical roof derives from the triangular shape made by the rafters and ceiling joists. The ceiling joists tie the exterior walls together, resisting the outward thrust on the exterior walls. Because the joists tie the rafters together as a unit, the

CATHEDRAL CEILINGS ARE TRICKY. The challenge boils down to keeping the walls from spreading during and after construction.

rafters carry the downward load on the ridge to the eave walls. Remove the joists, as with a cathedral ceiling, and two things happen. The rafters push out and bow the eave-wall plates, and the ridge becomes load bearing and sags because it isn't sized to bear a load.

Building cathedral ceilings means finding ways to duplicate the joist's function or eliminating the need for it, both during construction and as part of the permanent structure. Simply put, if you keep the bottoms of the rafters from spreading apart or if you keep the ridge from sagging, the roof will be strong and stable.

Proper bracing is the most important consideration during construction. If the walls aren't properly braced, the rafters' thrust will bow the plates and their weight will sag the ridge. And the wind, blowing against the tall gable-end walls typical of cathedral ceilings, can knock down the whole assembly.

I brace the eave and gable walls plumb at least every 8 ft. with 2×4s half again as long as the wall is high, or 12 ft. for an 8-ft. high wall. One-story gable walls call for two tiers of braces, one high and one low. Two-story gable walls get three tiers of braces, and I check the walls for plumb between each tier. I sometimes use 2×6s for the longest braces. The braces are nailed to the tops of studs and to 2-ft.-long 2×4 cleats that are nailed to a floor joist. I use at least two 12d nails at each end of a brace.

Straightening the top plates usually involves pulling the walls in as well as pushing them out. I pull them in with spring braces made from long 2×4s nailed on the flat to the top of the top plates and to the floor at a joist. Jamming a shorter 2×4 under the middle of a spring brace bends it and pulls in on the wall.

T-Shaped Posts Support the Ridge

If the ridge of a cathedral ceiling can be kept from sagging, the rafters can't push out the wall plates and the roof stays put. This is the principle behind the structural-ridge roofs that I will be describing later. With the other types of cathedral ceilings, though, the ridge board isn't load bearing except during

CEILING JOISTS HOLD A STANDARD ROOF TOGETHER

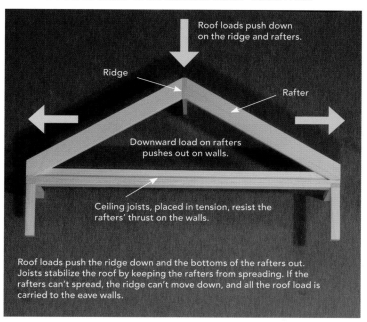

Roof loads push down on the ridge and rafters.

Ridge

Rafter

Downward load on rafters pushes out on walls.

Ceiling joists, placed in tension, resist the rafters' thrust on the walls.

Roof loads push the ridge down and the bottoms of the rafters out. Joists stabilize the roof by keeping the rafters from spreading. If the rafters can't spread, the ridge can't move down, and all the roof load is carried to the eave walls.

BRACING HOLDS WALLS PLUMB AND STEADY

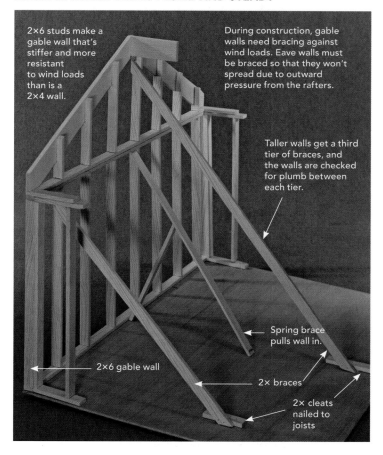

2×6 studs make a gable wall that's stiffer and more resistant to wind loads than is a 2×4 wall.

During construction, gable walls need bracing against wind loads. Eave walls must be braced so that they won't spread due to outward pressure from the rafters.

Taller walls get a third tier of braces, and the walls are checked for plumb between each tier.

Spring brace pulls wall in.

2×6 gable wall

2× braces

2× cleats nailed to joists

the construction process. Because of this, the ridge board is not sized to take a load and can sag from the weight of the rafters during construction.

To avoid this sagging, I support the ridge with T-shaped posts made by nailing a 2×4 on edge to the center of a 2×6. The T-shape of the post resists buckling under load better than does a single piece of lumber. I space the posts no more than 12 ft. apart and make sure that each one sits over well-supported floor joists. If I have doubts about a post sitting on one joist, I stand the post on a 2×10 or 2×12 laid flat over several joists to spread the load.

Rafters Must Make Room for Insulation

Many framers lay out roofs so that opposing rafters are staggered, making it easy to nail through the ridge into rafter ends. But it's usually best to align opposing rafters in a cathedral ceiling to allow subsequent members to be nailed evenly, instead of

at an angle. I keep this in mind when I lay out the mudsills so that floor joists and studs stack under the rafters.

Once I start framing and the customers can finally begin to see the house three-dimensionally, it's common for them to ask if a flat ceiling could become cathedral. If the rafters and ceiling joists aren't already cut, accommodating this request is usually a simple matter of stepping up the original rafter size at least one dimension. For example, I can use 2×8 rafters on a 28-ft. wide house that has a conventional roof. If the roof were changed to a cathedral style, I'd use at least 2×10 rafters. The reason is twofold. Extra heft helps to keep the rafters from sagging over time. And without flat ceiling joists, the insulation goes in the roof. The rafters must be wide enough to accommodate the insulation plus space for ventilation.

I have to be selective with rafter material when building cathedral ceilings. The underside of the rafters forms the ceiling plane, so any rafter material with extreme crowns that might show through the finish ceiling gets culled.

The bird's mouths in cathedral rafters have to be cut so that the bottom of the rafter intersects the corner of the top plate. If it doesn't, there would be an area between the ceiling plane and the wall with no nailing for drywall.

Hip and Valley Rafters Can't Hang below Commons and Jacks

Hip and valley rafters are often sized deeper than the rest of the rafters because they carry the combined loads of the jack rafters. Normally, nobody cares if a beefed-up hip or valley rafter hangs down below the other rafters into the attic. But with a cathedral ceiling, a deep rafter would protrude through the finished ceiling. Because of this, if your plans call for an oversize hip or valley rafter, they may have to be made from two smaller members nailed together. If you have any doubts about hip-rafter or valley-rafter size, consulting an engineer is wise. Alternatively, you could fur the ceiling out to the level of the protruding hip or valley. Finally, ventilation along a hip or valley requires some thought.

Once all the rafters are up, I usually sheathe the roof. The ridge is still supported with temporary posts, so the roof assembly is strong enough for my men to work on. Sheathing the roof at this point stiffens it and takes the bounce out of the rafters, making it easier to nail the subsequent members to them.

Raising the Ceiling Joists Is the Simplest Cathedral Ceiling

Raising the stable triangle of joists and rafters upward is not much more complex than framing a standard gable roof. It's probably the least expensive route, and the mix of angles and flats makes for an interesting ceiling. But if the triangle becomes too small, it can't stabilize the roof. I'll raise these joists about one-third of the distance from the top of the wall to the underside of the ridge. Lower is stronger.

Ceiling joists can often be raised higher than this, but a variety of factors comes into play. The room width, the roof pitch, and the snow load all must be considered. It's wise to consult a structural engineer before raising the joists higher.

I frame this roof much as I would a normal gable roof, starting with the end rafters, the gable walls, and the ridge. After supporting the ridge with a T-post, my crew sets the rafters.

After deciding their height, I install the joists. They must be level and in plane with each other. I measure up from the floor and mark the height on both gable walls. A joist is nailed at both ends of the room and checked for level.

I locate the rest of the joists with strings, rather than by snapping chalklines on the underside of the rafters. The rafters are never crowned exactly the same; thus, a chalkline won't be straight and the ceiling won't be flat. I cut blocks from a piece of scrap and nail them atop the ends of the gable joists. I string a line on each side of the room from these blocks and space the remaining joists down from the lines with other blocks. The joists don't touch the string, reducing the chance of accidentally

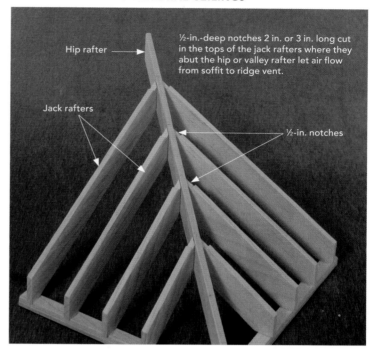

VENTING A HIP IN CATHEDRAL CEILINGS

Hip rafter

½-in.-deep notches 2 in. or 3 in. long cut in the tops of the jack rafters where they abut the hip or valley rafter let air flow from soffit to ridge vent.

Jack rafters

½-in. notches

pushing it out of line. The strings are set above the joists so that my crew doesn't have to wrestle them over the strings. Variation in joist width isn't usually a problem, particularly if all the stock comes from the same pile of lumber. The joists are nailed to the rafters with at least six 12d nails in each joint. I cut the joists to the roof angle so that there is more wood to nail into than if the joists were square-cut. I cut them just short enough so that they won't touch the roof sheathing. This way, the rafters won't shrink past the joist ends, creating bumps in the roof.

If the span is sizable, I use wider joists. For spans up to 14 ft., 2×6s are fine; beyond that, I increase to 2×8s. If the joists span more than 12 ft., I nail a 2×4 flat to the top of the joists, running perpendicular to the joists and centered in the span. A 2×6 on edge is nailed to the 2×4, creating a strongback. I place the strongback material on top of the joists before installing all of them. Otherwise, I won't be able to get the material up there at all.

A variation on this ceiling is to double the joists on every third rafter pair and leave out the interven-

RAISED CEILING JOISTS ARE THE SIMPLEST CATHEDRAL CEILING

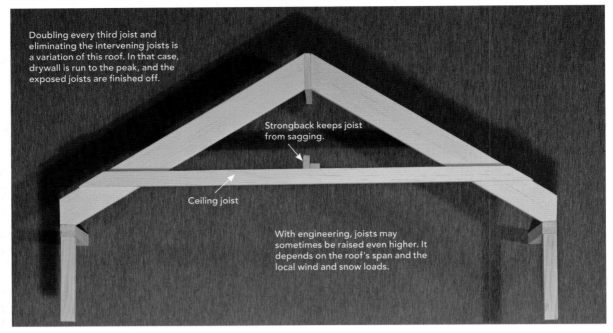

Doubling every third joist and eliminating the intervening joists is a variation of this roof. In that case, drywall is run to the peak, and the exposed joists are finished off.

Strongback keeps joist from sagging.

Ceiling joist

With engineering, joists may sometimes be raised even higher. It depends on the roof's span and the local wind and snow loads.

STRINGS HELP THE CARPENTERS TO ALIGN THE CEILING JOISTS

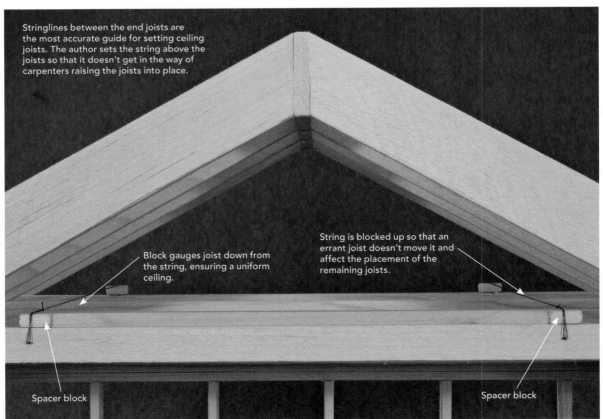

Stringlines between the end joists are the most accurate guide for setting ceiling joists. The author sets the string above the joists so that it doesn't get in the way of carpenters raising the joists into place.

Block gauges joist down from the string, ensuring a uniform ceiling.

String is blocked up so that an errant joist doesn't move it and affect the placement of the remaining joists.

Spacer block

Spacer block

ing joists. Similar caveats about not raising the joists more than one-third the roof height apply. On this ceiling, the drywall goes all the way to the peak. The doubled joists are exposed, and either drywalled or finished with trim stock.

Scissors Trusses Can Be Site-Built

A scissors truss consists of two opposing rafters braced by two pitched ceiling joists (or truss chords) that resemble lower-slope rafters. The chords cross at the ceiling's peak and continue upward to lap the rafters. This ceiling works well when the customer wants an unbroken ceiling plane right up to the peak. It's also good if the client wants the ceiling to be a shallower pitch than the roof is.

The chord's pitch shouldn't exceed two-thirds of the rafter's pitch. In other words, if the rafters are a 9-in-12 pitch, the chords should be a 6-in-12 or lower pitch. The steeper the pitch of the chords, the less effective they are at bracing the rafters. I make the chords one size smaller in depth than the rafters.

Framing a scissors-truss roof begins similarly to framing a raised-joist roof. Set the gables and the ridge. Brace the ridge, set the rafters, and partially sheathe the roof. Here, it's especially important to lay out the rafters so that they align at the ridge.

The gable rafters are supported by walls, so there is no need to brace them with chords. The gable-end chords essentially serve as drywall nailers and are nailed to the gable walls. I lay them out just like common rafters, without deducting for a ridge. After nailing up the gable-end chords, I cut the bird's mouth on a piece of chord stock that's long enough to span from the wall to the opposing rafter. I hold this chord stock in place, even with one of the gable-

SCISSORS TRUSSES REDUCE THE CEILING PITCH

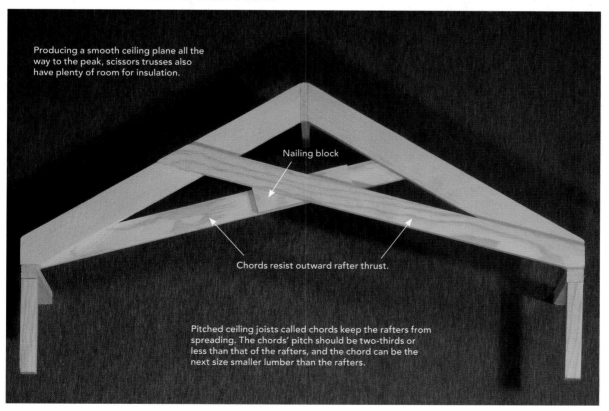

Producing a smooth ceiling plane all the way to the peak, scissors trusses also have plenty of room for insulation.

Nailing block

Chords resist outward rafter thrust.

Pitched ceiling joists called chords keep the rafters from spreading. The chords' pitch should be two-thirds or less than that of the rafters, and the chord can be the next size smaller lumber than the rafters.

STRUCTURAL RIDGE BEAM KEEPS ROOF FROM SAGGING

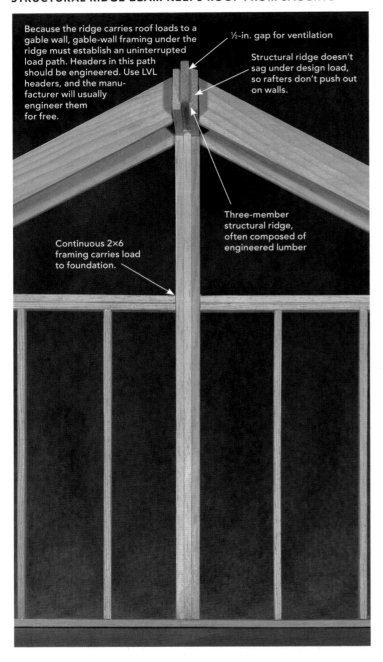

Because the ridge carries roof loads to a gable wall, gable-wall framing under the ridge must establish an uninterrupted load path. Headers in this path should be engineered. Use LVL headers, and the manu-facturer will usually engineer them for free.

½-in. gap for ventilation

Structural ridge doesn't sag under design load, so rafters don't push out on walls.

Three-member structural ridge, often composed of engineered lumber

Continuous 2×6 framing carries load to foundation.

sides of rafter pairs with six 12d nails per joint. I also toenail them to the wall plate. The chords are lined up on the strings and nailed to the rafter on the far side of the ridge. Where the chords cross, they're the thickness of the rafter apart. I nail a 2-ft. block of the chord material flush with the bottom of one chord and nail the second chord to the block.

Design the Ridge as a Beam, and No Joists Are Needed

Another approach to cathedral ceilings is to make the ridge a beam that's stiff enough not to sag under load. I build this type of roof when a ceiling that climbs cleanly to the peak at the roof pitch is wanted.

A structural ridge creates point loads that must be carried through the gable wall to the foundation with continuous, stacking framing. Headers in this load path need to be sized accordingly, and their studs may need beefing up, too.

I balloon-frame particularly tall gable walls. The studs in balloon-framed gable walls reach from the bottom plate on the first floor to the top plate just below the rafters. Balloon-framing avoids the plates at the various floor levels common to platform-framed walls. Plates can act as a hinge, weakening tall walls. To stop the chimney effect these continu-ous stud cavities can have in a fire, codes specify fire blocking at least every 8 ft. and where the wall intersects floors and ceilings.

In areas where gable walls are subject to high wind loads, I frame gable walls with continuous 2×6 LSL studs. LSL is factory made by shredding lumber and gluing the strands back together. LSL is denser and stiffer than solid-sawn lumber, and makes for a stronger but more expensive wall.

I avoid large, single-member ridge beams. They're heavy and often must be placed with a crane. I prefer to assemble in place two, three, or even four full-length LVL members that can be lifted by hand. LVLs can span greater distances than standard lumber and are made from material similar to LSL studs. LVL manufacturers will usually size the beam for you at no extra cost.

end chords. By marking the chord stock where it laps the opposing rafter, I have the pattern for the rest of the chords.

To line up the chords, I string two lines from the top of the end chords, just as I did with the raised-ceiling-joist roof. The chords are nailed on opposing

BEAM UNDER RIDGE CARRIES ROOF LOAD

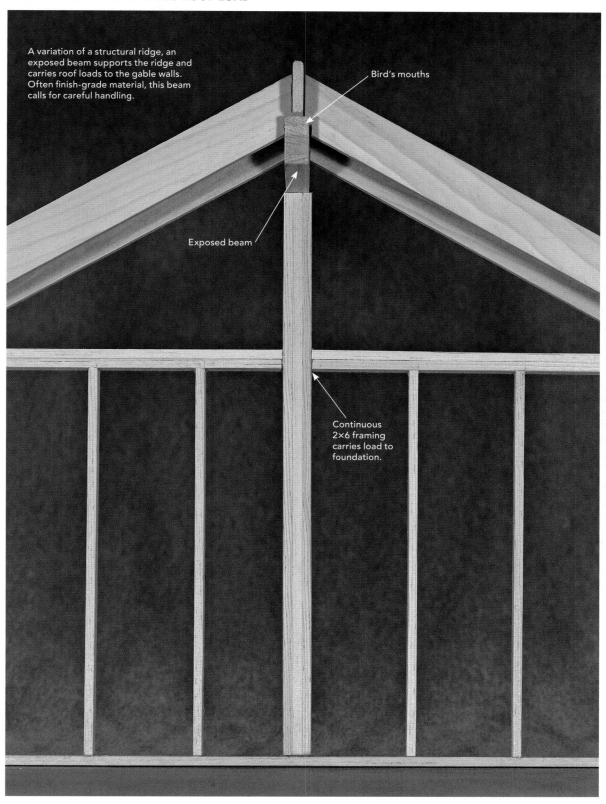

A variation of a structural ridge, an exposed beam supports the ridge and carries roof loads to the gable walls. Often finish-grade material, this beam calls for careful handling.

Bird's mouths

Exposed beam

Continuous 2×6 framing carries load to foundation.

Individual ridge members longer than 24 ft. are usually too heavy to lift by hand. In that case, I'll assemble the beam on the ground and lift it with a crane. It's important to build beams straight; once nailed, they're nearly impossible to straighten.

Before nailing together a multimember beam in the air, I set the gable rafters and wall. On 2×6 gable walls, my crew sets the first beam member between the gable rafters just like a ridge board. The subsequent members are cut shorter so that they butt to the inside of the gable rafters. Even after deducting 1½ in. for the gable rafter, the beam has a full 4 in. of bearing. I stagger these shorter members down so that they're about ½ in. lower than the top of the rafters. This method allows air to flow to the ridge vent. After nailing the beam together, I measure, cut, and then install the post under the beam to carry the load downward.

For 2×4 gable walls, all members of the ridge beam must run through the entire width of the wall to gain sufficient bearing. This means that all the beam members have to be placed at the same time as the gable rafters, a trickier operation. Because of this situation and because 2×6 gable walls are stiffer, I rarely build 2×4 gable walls when using a structural ridge.

Show a Finished Beam Beneath the Ridge

This roof goes up similarly to the previous example, except that the beam is installed below and supports a standard ridge. I build this type of roof when the customer wants to show a large finished beam or when the ridge beam is so deep that it would hang below the rafters anyway. In that case, I often put collar ties just below the beam for drywall nailers. This eliminates the need to drywall and finish that awkward triangular space between the rafters and the side of the beam.

Shorter beams that are light enough to be handled by a couple of carpenters can be installed after the rafters are set. With longer beams, however, especially big single-member beams, it's easier to set the beam first, then build the roof around it.

Again, the first step is building the gable walls and setting the gable rafters. The wall must have a post to support the beam, just as in the structural-ridge type of roof. I cut the gable rafters normally and set the beam within them by hand or by crane.

If this beam is to show, I treat it with care. I hoist it with nylon slings instead of chains, which can mar the surface. And I don't nail temporary braces to the finish face. The nail holes might show, and worse, if the nails rust, they'll deeply stain the beam.

The rafter tops will have bird's mouth cuts in them that fit over the beam. I don't toenail through the upper seat cut; this usually splits the top of the rafter. Rather, I nail the rafter to the ridge and toenail the ridge to the beam. When laying out this seat cut, I allow for the height of the ridge board plus ¼ in. or so. The rafters don't have to touch the beam because the ridge does. This ¼ in. allows a bit of play that simplifies setting the rafters.

Open Up the Ceiling with a Steel Sandwich

BY MICHAEL CHANDLER

My company specializes in designing and building small homes. I like to vault ceilings in small rooms—bedrooms and screened porches, commonly—that normally would have flat ceilings.

People love cathedral ceilings because they add drama and presence to interior spaces. A vaulted ceiling can make a small room look large rather than cramped and confined. So why are cathedral ceilings usually reserved for living rooms or other large public areas, whereas small rooms get stuck with flat ceilings? Cost.

Scissors trusses or structural ridge beams are the common, expensive methods for incorporating cathedral ceilings in houses. For the past few years, though, my company has overcome this economic problem by using angled steel flitch plates supplied by a steel fabricator. As shown in the illustration on p. 225, the V-shaped plate is angled to match the roof peak, and it's sandwiched between common rafters. Angled steel flitch plates work well for pyramid roofs but can be used to open gable roofs, too. But keep in mind that any steel beam in a home will be outside the prescriptive guidelines of code and will need to be reviewed by a local engineer in the context of the specific building load paths.

Scissors Trusses Don't Cut It

For years, I've been building vaulted ceilings using scissors-truss kits from my local truss supplier, but scissors trusses have a number of drawbacks: less space for insulation at the top plate, scheduling headaches, and cost. Did I mention that they're also a pain in the neck to install? Getting the little trusses to line up for a smooth roof and ceiling is fussy at best; my crew and I often have to strap the ceilings with 1×4s to get them smooth enough for drywall. The truss package for the last pyramid hip ceiling I built, for a 16-ft. by 16-ft. bedroom, cost $1,600*. The three-week lead time added insult to injury. There had to be a better way.

My engineer came up with an angled flitch-plate detail that creates a V-shaped beam capable of supporting 2×12 framing and that adds only $180 to the framing costs. It's like a 160-lb. Simpson Strong Tie. The welding must be done by an American Welding Society certified welder using a full-penetration weld; you can't just do it on site. The first time I brought one of these angled-flitch drawings to my welder, he wanted a plywood template. Now I can fax him a diagram and he has the flitch plate ready for me to pick up in a week, complete with clean holes.

BOLTED BETWEEN RAFTERS, an angled steel flitch plate can eliminate the need for rafter ties in cathedral ceilings.

ENGINEERING A CATHEDRAL CEILING

THE WEIGHT OF A ROOF NATURALLY CAUSES rafters to thrust outward, which can push the walls apart. In a typical house ceiling, joists prevent the walls from spreading. In a cathedral ceiling, there are a few common ways to offset this outward thrust.

- **A structural ridge** is a beam large enough to span the length of the house and support the weight of the roof.

- **Rafter ties** provide tension force, which offsets the outward thrust and works for hip or gable roofs. The downside: horizontal framing members.

- **Scissors trusses** are engineered to distribute forces internally on hip or gable roofs. The downsides: reduced ceiling pitch, difficult to drywall on hip roofs, expensive.

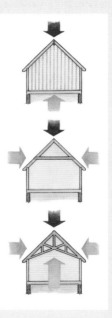

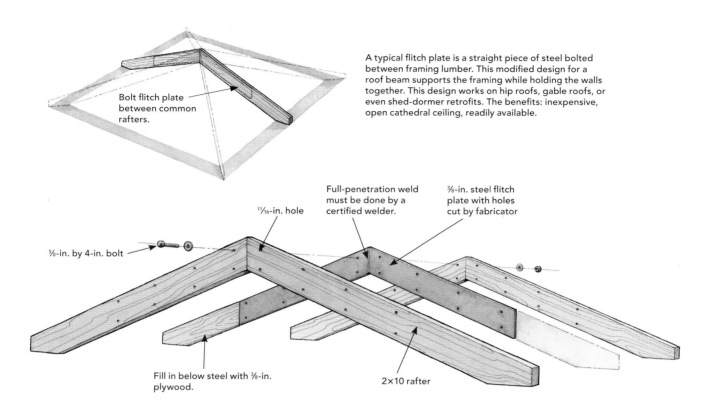

Bolt flitch plate between common rafters.

A typical flitch plate is a straight piece of steel bolted between framing lumber. This modified design for a roof beam supports the framing while holding the walls together. This design works on hip roofs, gable roofs, or even shed-dormer retrofits. The benefits: inexpensive, open cathedral ceiling, readily available.

Full-penetration weld must be done by a certified welder.

³⁄₈-in. steel flitch plate with holes cut by fabricator

1¹⁄₁₆-in. hole

⁵⁄₈-in. by 4-in. bolt

Fill in below steel with ³⁄₈-in. plywood.

2×10 rafter

This flitch solves all the problems of scissors trusses: It has a full 10 in. for insulation over the top plate, it keeps the ceiling the same pitch as the roof, and it ensures that the roof and ceiling planes are smooth every time. Yes, the steel is heavy, and getting all those holes to line up takes some planning. But the assembled flitch beam is much less awkward to handle than scissors trusses, and it costs about $1,000 less for that 16-ft. by 16-ft. room.

The Holes Are Bigger Than the Bolts

The holes in the steel should be cut by the fabricator, preferably with a high-pressure hydraulic "piranha" punch. Do not blast holes in the steel with a cutting torch because the holes will be ragged and will catch at the drill bit while you're using the steel as a template to drill holes into the wood.

GET HELP. Installing scissors trusses requires one or two helpers.

When drilling through the rafters, use the same-size drill bit as the hole in the steel. This step ensures good contact for the bolt between the wood and the steel. The bolt, however, should be slightly smaller than the hole. An ¹¹⁄₁₆-in. hole in the steel requires a ⅝-in. bolt. Undersizing the bolt more than that weakens the connection because the steel can move in the sandwich; undersizing the holes in the rafter can cause the wood to split when the bolts are tightened.

Assemble the Beam in Place

Because the flitch weighs a lot (160 lb., in this case), the installation is not a one-person job. You'll need a helper or two to build this beam in place. I set up a temporary catwalk down the center of the room with a support post underneath. The catwalk needs to be stiff enough to support the framing lumber, the steel, a helper, and me.

TACK THE RAFTERS FLAT. After laying the rafters atop the wall plates and a temporary catwalk, nail the plumb cuts tightly together. Align the seat cuts to their layout marks on the wall plates, and toenail.

THE STEEL IS SHORTER THAN THE RAFTERS. Fill in the ends of the steel plate with ⅜-in. plywood and 1-in. roofing nails. After the top rafter is bolted in place, 8d nails driven from each side will hold the tails together.

USE THE STEEL AS A DRILLING TEMPLATE. After laying the steel in place on top of the rafters, bore a hole at each end of the steel, then insert a bolt to hold the steel in place while the remaining holes are drilled.

COMPLETE THE SANDWICH. Clamp the top rafters in place and drill up through the previously drilled holes. Again, make sure the plumb cuts are tight before drilling the holes.

With the catwalk ready, I set up my sandwich shop. I lay a couple of common rafters flat and tack them in place on the catwalk and wall tops. I tack the rafters to the wall plates so that the nail will act as a hinge; that is, the nail will hold the rafter in place during assembly and keep it from sliding when it's time to stand up the beam.

Next, I lay the steel on the rafters and drill down. To prevent the plate from shifting around, I drop several bolts into holes as I go. Having a helper comes in handy for more than just lifting the steel. A helper minimizes my number of trips up and down the ladder for bolts, nuts, washers, and plywood filler strips. Have a helper cut and fit ⅜-in. plywood spacers as you drill holes.

With all the holes drilled in the lower-rafter pair, I pull the bolts (except for one in the tail of each rafter), lay the second rafter pair in position, and clamp them in place. The remaining bolts keep the parts aligned while the first couple of holes are drilled in the second rafter pair. Drill up from below at the peak, and set bolts in. After removing the first pair of bolts, I snug the second pair down to prevent drill vibration from causing the parts to lose alignment.

Stand and Brace the Beam

Once all the bolts are tightened down, tilt the assembly upright and brace it well. It's not that heavy as long as the ends are well secured on the top plates.

A NAIL HOLDS THE BEAM WHILE IT'S TILTED UP

PLACE THE FIRST RAFTER EXACTLY ON THE layout line so that when it's rolled up the rafter will be in the right place. Drive a 16d nail at an angle through the rafter and into the top plate. As the beam is lifted, the nail bends but keeps the rafter on its layout line. Have 2×4 braces on hand to nail into the beam after it's upright.

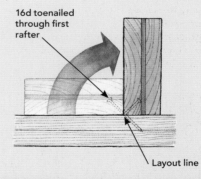

16d toenailed through first rafter

Layout line

Have the bracing ready to nail as soon as the assembly is plumbed. Although 2×4 braces work fine, I like to set a perpendicular common rafter and a couple of hip rafters as bracing if I'm unable to finish the roof framing that day.

The rest is just regular hip-roof stick-framing, except that the rafters are loaded in shear at the peak, so you need to use a hanger designed for shear load.

This type of roof assembly requires more uplift reinforcement than a comparable roof with ceiling joists that are used as collar ties, because there's less meat for toenailing into the top plate. For this reason, exceeding the code requirement for rafter tie-downs is a great idea.

* Prices are from 2006.

USE STEEL FRAMING CONNECTORS FOR SHEAR RESISTANCE. The steel flitch-beam assembly shoulders the shear load in one direction, but the intersecting hip rafters should be supported with a special framing anchor (the Simpson HRC22 hip-rafter connector) as well.

PART 6

Energy-Efficient Framing

6 Proven Ways to Build Energy-Smart Walls

BY BRUCE COLDHAM, FAIA

If we look even a mere 20 years into the future, we'll see a very different world. For one thing, fossil-fuel-based energy will be much less abundant and much more expensive as we become independent of foreign oil. Eventually, we will be living in a world without oil. But the places we live in last a lot longer than 20 years, so we should start thinking now about how to construct houses that will function in a post-petroleum economy.

To that end, a good deal of money and brainpower is being put to work to come up with efficient solutions. The German Passive House Institute and its U.S. affiliate (Passive House Institute US; www .phius.org) have designed building-enclosure standards based on energy use that's 15 percent to 20 percent of today's typical residential design load. Their model suggests that an envelope ratio of 5:10:20:40:60 should be considered a minimum in the northern tier of the United States. In other words, windows at R-5, subslab at R-10, basement wall or slab perimeter at R-20, walls at R-40, and ceiling or roof at R-60.

Fortunately, you can achieve this goal now with one of six high-performing wall systems, each capable of achieving or exceeding R-40 (not including the added values of the siding, the drywall, or the surface-bound air layers). Each can be interpreted or varied in many ways. Although there are still other approaches to consider, I'm focusing on systems currently in use (or in development) in the northern United States, where the climate demands more efficiency from a structure.

Airtightness, Moisture Storage, and Drying Potential Are Key

Air leaks are a wood-framed wall's worst enemy. In cold-climate, high-R buildings, the exterior sheathing spends a significant portion of the year below the dew point of the interior air because of heavily insulated walls. If warm interior air leaks out through the wall cavity, the moisture it carries will condense on the sheathing and can result in rot and mold growth. Great care must be given to achieve less than one air change per hour at 50 pascals (1 ACH50), or less than 0.1 cfm of air leakage per square foot of shell area. This is five times tighter than the Energy Star airtightness standard and creates the need for mechanical ventilation.

In each of the following wall-system examples, the air barrier must extend unbroken up and across the roof or ceiling, and down to the basement wall or floor slab.

STANDARD R-15 WALL
R-Value 15.0

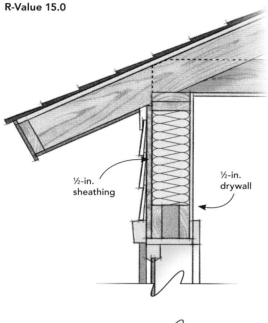

½-in. sheathing

½-in. drywall

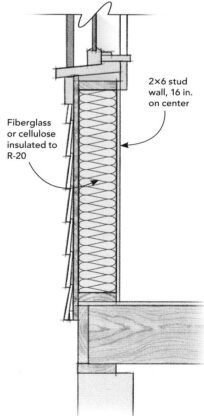

Fiberglass or cellulose insulated to R-20

2×6 stud wall, 16 in. on center

There also can be no water leaks in high-efficiency walls. Windows should be installed on full sill flashing and have wrapped jambs. If the wall does get wet, it should be able to store moisture, then gradually dry either to the interior or exterior and endure for decades. A wall must lose rather than gain water over the course of a year. Of course, it's best that the wall not get wet in the first place. But think of such recuperative capability as a form of insurance that is designed and built-in rather than purchased afterward.

Cost Is Relative

An energy-efficient wall system should contribute to lower heating and cooling costs. It also should reduce energy loads on mechanical systems so that smaller, less expensive heating and/or cooling units can be integrated into a house. In some cases, a central-heating system can be eliminated, and the house can rely on point-source space heaters or single-port air-source heat pump units. Beyond cutting down on energy consumption, a well-built wall system reduces drafts and maintains interior relative-humidity levels, which increases overall comfort, while the absence of thermal bridges eliminates those cold surfaces where moisture condenses and mold can grow, making for healthier living.

Of course, all this comes at a cost, which is substantially higher than if the walls are built to code minimums. Finding this balance is what makes the cost equation difficult. Even among the six wall systems shown here, useful cost comparisons are hard to make because they depend on contractor familiarity and on regional markets. I think one thing is ultimately true: In smaller, open-plan houses where these high-performance walls will permit eliminating a central-heating system, the cost premium will be retrieved, even without staggering fuel-price escalations.

Double Stud Walls Are Tried and True

Double walls have been built since the early days of superinsulated houses. They typically consist of a conventionally framed 2×4 exterior wall sheathed with ½-in. structural panels. A second non-load-bearing 2×4 stud wall is built on the interior. Half-inch or ¾-in. plywood boxes span the door and window rough openings between the inner and outer frames. At 11¾ in. deep (an economical cut from a 4-ft. panel), these boxes determine the exact width of the wall, which leaves a 4¼-in.-thick zone of uninterrupted insulation in the wall's center.

In previous projects, my firm has used dense-pack cellulose in the cavity after the electrical rough-in is complete. (Adding high-performance fiberglass will increase the R-value.) Although electrical wiring is anticipated within the cavity, we keep all HVAC ducting out of exterior walls.

Blowing cellulose insulation into these deep cavities requires some planning. We either install

DOUBLE STUD WALLS

R-value 40.0

SUITABLE FOR:
- New construction
- Retrofit

THERMAL PERFORMANCE
At R-3.5 per in., the cellulose in this system yields an overall value of R-40.

PROS
- Familiar building process means lower labor costs with readily available materials.
- If cellulose is used, the project has a lower environmental impact.
- Large moisture-storage capacity allows wall cavities to dry gradually.

CONS
- Fairly involved insulation process can add to labor costs.
- Reduces interior space by 3 percent
- Exterior portion of wood framing sees very cold conditions and is vulnerable if air sealing is lax and/or if interior relative humidity (RH) is constantly high.

vertical netting between the inner and outer studs to define each bay and fill each separately, or we adopt the two-stage process of a bulk blow topped off with a smaller-diameter hose to achieve the uniform 3.5 pcf (pounds per cubic foot) density. No vapor retarder is used except for the interior paint.

Larsen Trusses Create an Exoskeleton

Like the double stud wall, a Larsen truss wall creates an airspace. A load-bearing 2×4 frame wall is sheathed with ½-in. plywood or OSB. An 8-in.-deep exoskeleton frame is attached to the load-bearing wall and is supported on a wide foundation wall.

DOUBLE STUD WALLS

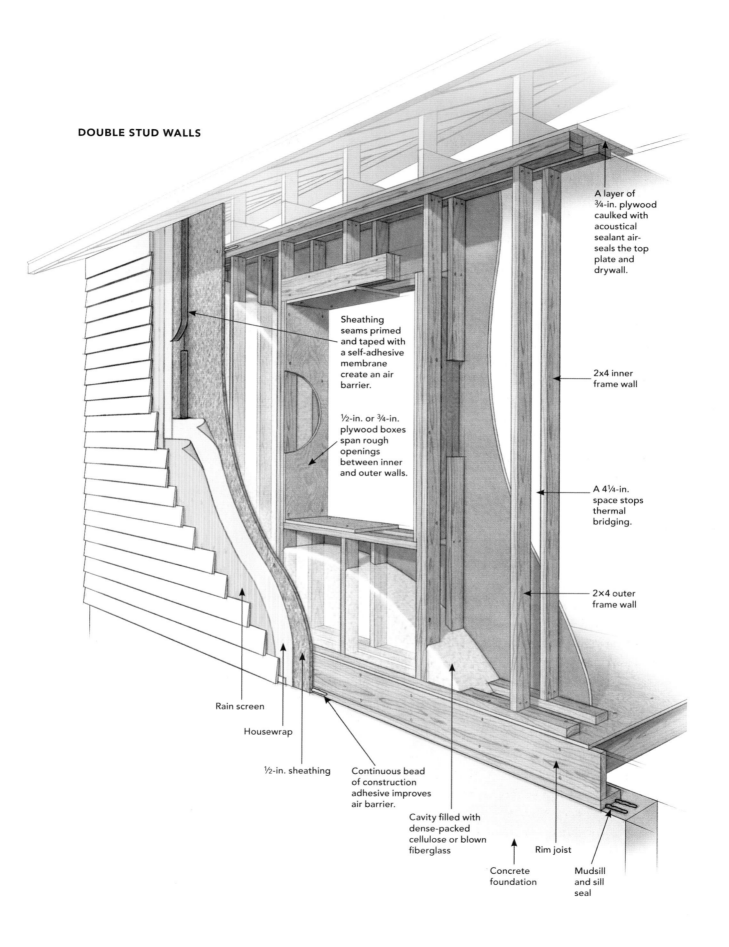

A layer of ¾-in. plywood caulked with acoustical sealant air-seals the top plate and drywall.

Sheathing seams primed and taped with a self-adhesive membrane create an air barrier.

½-in. or ¾-in. plywood boxes span rough openings between inner and outer walls.

2x4 inner frame wall

A 4¼-in. space stops thermal bridging.

2×4 outer frame wall

Rain screen

Housewrap

½-in. sheathing

Continuous bead of construction adhesive improves air barrier.

Cavity filled with dense-packed cellulose or blown fiberglass

Rim joist

Concrete foundation

Mudsill and sill seal

LARSEN TRUSSES

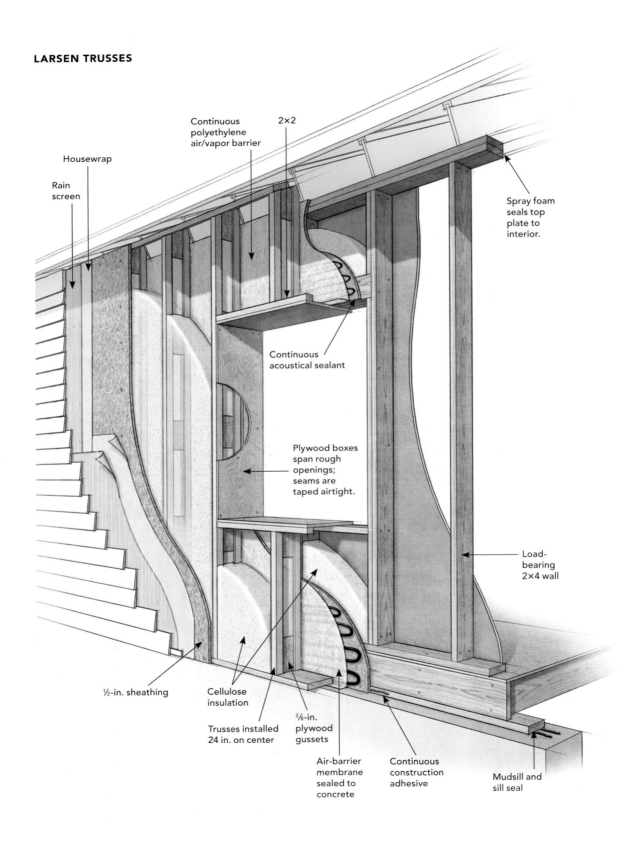

Rain screen

Housewrap

Continuous polyethylene air/vapor barrier

2×2

Spray foam seals top plate to interior.

Continuous acoustical sealant

Plywood boxes span rough openings; seams are taped airtight.

Load-bearing 2×4 wall

½-in. sheathing

Cellulose insulation

Trusses installed 24 in. on center

⅜-in. plywood gussets

Air-barrier membrane sealed to concrete

Continuous construction adhesive

Mudsill and sill seal

R-value 39.5

SUITABLE FOR:
- New construction
- Retrofit

THERMAL PERFORMANCE
Used in this system, cellulose (at R-3.5 per in.) yields an overall value of R-39.5.

PROS
- Continuously protected air barrier keeps wall cavity dry.
- Large moisture-storage capacity allows wall to dry gradually.
- During retrofits, interior walls can remain intact.

CONS
- Unconventional framing can increase production time and labor costs.
- Increases the gross building area by 3 percent.
- Exterior portion of wood framing sees very cold conditions and is vulnerable if air sealing is lax and/or if interior RH is constantly high.

The floor deck and bearing wall are held to the inner half of the foundation wall. Both are insulated in retrofits. The truss exterior is not supported by the foundation and is sealed with a plywood bottom plate. A plastic air barrier and a vapor-retarding membrane are wrapped around the entire outside of the sheathed 2×4 frame and are sealed to the sill plate or directly to the concrete foundation wall. The assembly is made to dry both to the outside and the inside. The outer Larsen truss wall is typically made on site and creates an uninterrupted cavity filled with dense-pack cellulose. The insulation is blown through holes in a second exterior layer of sheathing.

Because the double-wall system is easier to air-seal, we primarily use the Larsen truss as a retrofit option. As with the double-wall system, ½-in.- or ¾-in.-deep plywood boxes are fabricated to span the rough openings, and here they also provide an additional measure of rigidity to the exoskeletal framing. The 2×4 wall inside the air barrier can be penetrated with utilities without compromising the airtightness of the house.

Inspired by the Passive House, Walls Made of Engineered Lumber

This system was used by Katrin Klingenberg to introduce the German Passive House program to the United States. Klingenberg's walls are built with 12-in.-wide I-joists as studs. The interior sides of the I-joists are sheathed with OSB that serves as the primary air barrier. The exterior of the I-joists is sheathed with structural fiberboard, which is vapor permeable and allows the wall cavity to dry to the exterior. These sheathings also give the I-joist wall a box-beam-like structural integrity. It pays to specify the window placement and widths so that rough openings fall between the regular stud spacing. The wall is supported on a concrete slab.

Walls can be site-constructed as panels and tipped up with the help of a crane. However, Klingenberg prefers to prefabricate double-story panels of about 8 ft. in width and to fill the wall cavities with high-efficiency blown fiberglass (R-4.5 per in.). (Cellulose also can be used.) The panel junctions are designed with an overlap that is glued and then taped to ensure that the enclosure stays airtight as it is assembled.

If you're pursuing the Passive House standard, a 2×3 wall built on the interior of the wall panels and insulated with damp-spray cellulose can support the second-floor joists and position the electrical and plumbing inside the air barrier. Having the vapor-retarding OSB sheathing slightly insulated from the interior-wall surface helps to eliminate moisture issues during the cooling season.

WALLS MADE OF ENGINEERED LUMBER

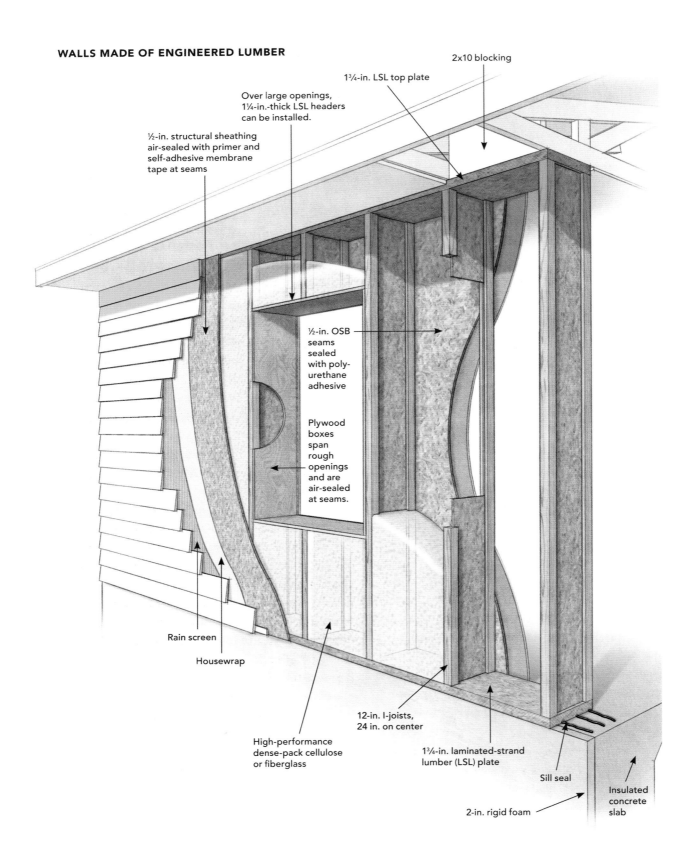

Over large openings, 1¼-in.-thick LSL headers can be installed.

1¾-in. LSL top plate

2x10 blocking

½-in. structural sheathing air-sealed with primer and self-adhesive membrane tape at seams

½-in. OSB seams sealed with poly-urethane adhesive

Plywood boxes span rough openings and are air-sealed at seams.

Rain screen

Housewrap

12-in. I-joists, 24 in. on center

High-performance dense-pack cellulose or fiberglass

1¾-in. laminated-strand lumber (LSL) plate

Sill seal

Insulated concrete slab

2-in. rigid foam

A Spray-Foam Shell Encases the House

This approach is particularly well suited to a deep-energy retrofit. Developed by John Straube, a building scientist and construction consultant based in Waterloo, Ontario, the system consists of vertical 2×3 furring spaced off existing exterior walls, then set in place with closed-cell spray foam. According to Straube, the furring needs only a minimum of attachment strength and stiffness.

The key to this system's strength is the closed-cell polyurethane foam sprayed around the furring. The density of the cured foam is sufficient to stabilize the furring and carry the load of the cladding. Indeed, beyond stabilizing the furring, the foam provides a continuous insulation blanket and an air barrier around the building.

The sprayed foam is installed to spill up over the existing wall's top plate to seal against the top side of the ceiling drywall, which forms the air barrier. The lap siding is attached directly to furring substantially embedded in the foam. In turn, the foam is applied so that its cured surface is recessed ½ in. to ¾ in. below the face of the furring, leaving an airspace behind the siding, which is vented top and bottom.

A Double Layer of Rigid-Foam Panels Blankets the Exterior

Conceptually the most simple of the six systems, this technique is particularly well-suited to retrofits or for enhancing conventionally framed new construction. In an approach developed by architect Betsy Pettit of Building Science Corporation, a double layer of 2-in. polyisocyanurate insulation is applied directly to the studs and is held in place with vertical 1×4s screwed to the framing.

With this system, the drainage plane and the air barrier can be either in front of or behind the rigid insulation. The decision largely depends on the windows and, specifically in a retrofit, whether they are kept or replaced. If the existing windows are kept, the drainage plane (and therefore the air-barrier membrane) should be installed on the exterior of the existing sheathing before the insulation is applied. This detail eliminates the need to tape the joints of the insulation panels.

For new construction (or for a retrofit that involves window replacement), the windows are best located in the plane of the applied 1×4 furring. The drainage plane is created by taping the insulation-panel joints. The air barrier can be established at the exterior, or it can remain behind the insulation, either as a sealed membrane or as taped joints of the wall sheathing.

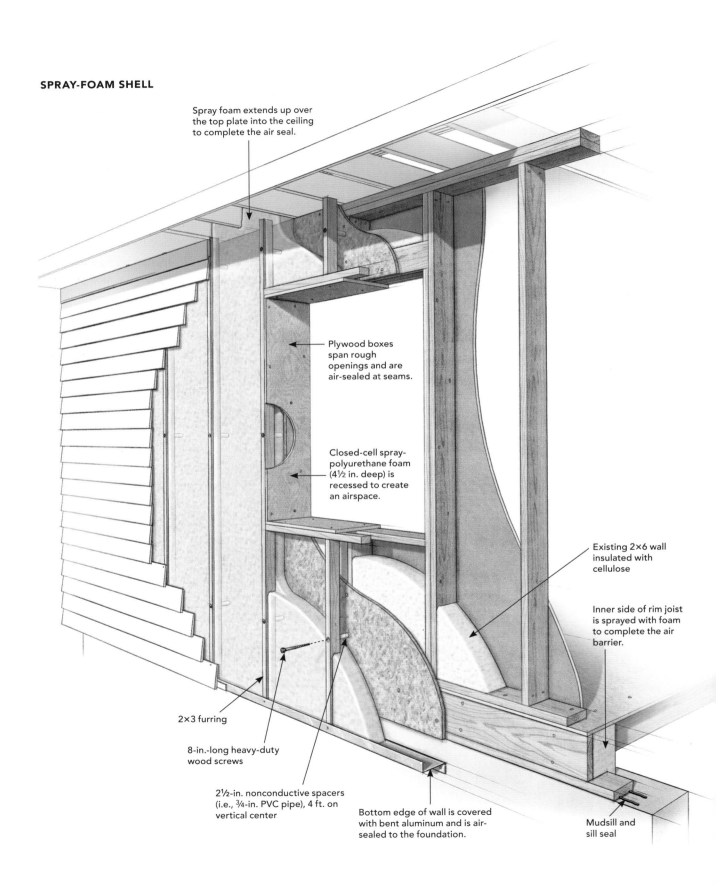

SPRAY-FOAM SHELL

Spray foam extends up over the top plate into the ceiling to complete the air seal.

Plywood boxes span rough openings and are air-sealed at seams.

Closed-cell spray-polyurethane foam (4½ in. deep) is recessed to create an airspace.

Existing 2×6 wall insulated with cellulose

Inner side of rim joist is sprayed with foam to complete the air barrier.

2×3 furring

8-in.-long heavy-duty wood screws

2½-in. nonconductive spacers (i.e., ¾-in. PVC pipe), 4 ft. on vertical center

Bottom edge of wall is covered with bent aluminum and is air-sealed to the foundation.

Mudsill and sill seal

R-value 41.5

SUITABLE FOR:
- Retrofit

THERMAL PERFORMANCE
A combination of 4.5 in. of exterior spray foam at R-5.6 per in. over a 2×6 interior frame, 24 in. on center, and filled with cellulose at R-3.5 per in., yields a value of R-41.5.

PROS
- Easily applied to retrofits of many building types.
- Cured foam forms an air and vapor barrier and rain-screen wall.
- All wood framing is warm and remains above any conceivable dew point temperature.

CONS
- Spraying conditions are weather dependent and could affect scheduling.
- Unusual technique can create higher labor costs.

Structural Insulated Panels (SIPs)

This wall-panel system consists of rigid-foam insulation sandwiched between two skins of $\frac{7}{16}$-in.-thick OSB. The best-performing foam core is polyisocyanurate. Chemical additives—in particular the brominated fire retardant HBCD—in polystyrene (EPS and XPS) make polyiso a better choice from a human- and environmental-health standpoint. Polyiso foam is also a more resourceful choice. The panels are typically constructed on a custom basis with rough openings blocked out, which reduces the waste that accrues from blank XPS core panels cut on site.

The OSB facings are particularly susceptible to degradation if they get wet. Therefore, SIPs require special attention during construction to keep them dry. It's also extremely important to air-seal the panel joints with face-applied SIP tape to prevent moisture-laden interior air from reaching the exterior skin.

Joe Lstiburek of Building Science Corp. says, "The best way to deal with the limited (almost nonexistent) moisture-storage capacity of SIPs is to ventilate exterior claddings and to add a vented 'overroof.' With those two additions, SIPs becomes pretty indestructible."

DOUBLE LAYER OF RIGID-FOAM

R-value 40.0

SUITABLE FOR:
- New construction
- Retrofit

THERMAL PERFORMANCE
Rigid foam: R-6 per in., 4 in. over 2×6 stud frame filled with cellulose; R-3.5 per in., R-40.0.

PROS
- Readily available materials.
- Placement of air barrier and drainage plane is adaptable to the situation.
- Relatively low-tech system can reduce labor costs.
- All wood framing is warm and remains above any conceivable dew point temperature.

CONS
- Modest moisture-storage capacity.

DOUBLE LAYER OF RIGID-FOAM

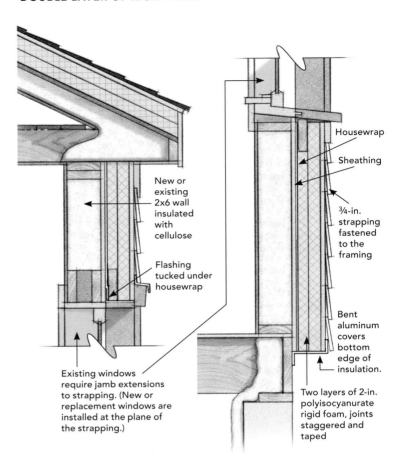

New or existing 2x6 wall insulated with cellulose

Flashing tucked under housewrap

Existing windows require jamb extensions to strapping. (New or replacement windows are installed at the plane of the strapping.)

Housewrap

Sheathing

¾-in. strapping fastened to the framing

Bent aluminum covers bottom edge of insulation.

Two layers of 2-in. polyisocyanurate rigid foam, joints staggered and taped

STRUCTURAL INSULATED PANELS (SIPS)

R-value 40.0

SUITABLE FOR:
- New construction

THERMAL PERFORMANCE
A core of 7³⁄₈-in.-thick XPS (R-5 per in.) yields R-38. A core of 5½-in.-thick polyisocyanurate (R-5.8 per in.) yields R-40.

PROS
- Once on site, panels assemble very quickly.
- Typically thinner wall profiles.

CONS
- Polystyrene's byproducts include toxins such as HBCD, benzene, and toluene.
- Vulnerable to moisture degradation.
- Electrical installation can be tricky.

STRUCTURAL INSULATED PANELS (SIPS)

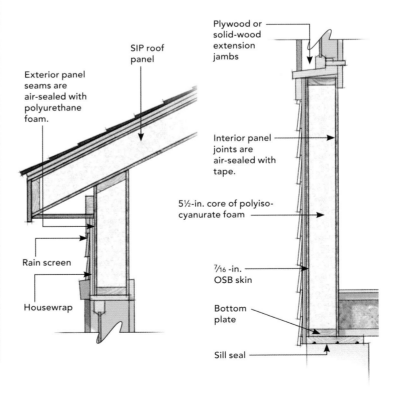

Exterior panel seams are air-sealed with polyurethane foam.

SIP roof panel

Rain screen

Housewrap

Plywood or solid-wood extension jambs

Interior panel joints are air-sealed with tape.

5½-in. core of polyiso-cyanurate foam

⁷⁄₁₆-in. OSB skin

Bottom plate

Sill seal

Air-Sealed Mudsill Assembly

BY STEVE BACZEK

The mudsill is one of the most critical components of a successful Passive House. It involves a connection between dissimilar materials, and making such a connection airtight is a challenge. Even the best stemwall will have some imperfections. Also, the mudsill typically will be wet from its preservative treatment and from the lumberyard, and it will shrink as it dries. This means that there likely will be gaps between the wood and the concrete. Traditionally, this part of the building is sealed with a foam gasket. In a Passive House, however, even a minor gap is a major problem, so our assembly is a bit more complex.

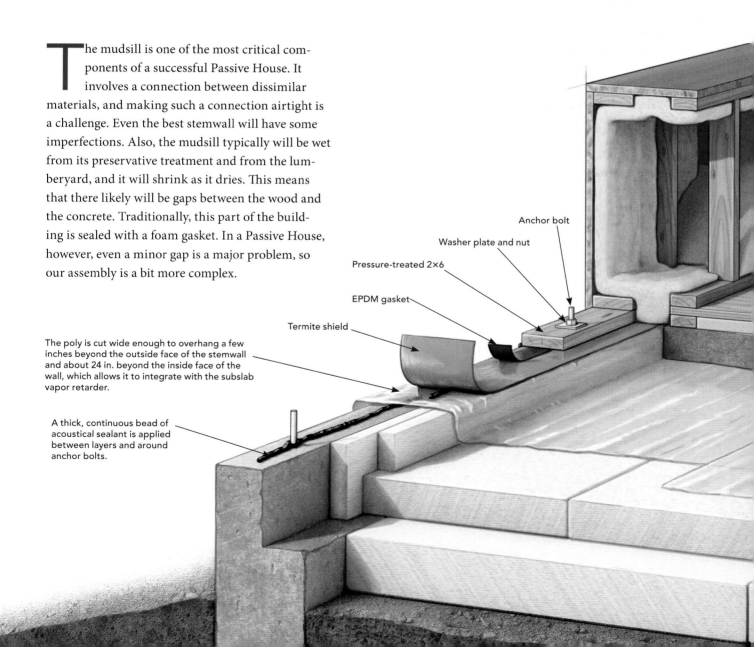

Anchor bolt

Washer plate and nut

Pressure-treated 2×6

EPDM gasket

Termite shield

The poly is cut wide enough to overhang a few inches beyond the outside face of the stemwall and about 24 in. beyond the inside face of the wall, which allows it to integrate with the subslab vapor retarder.

A thick, continuous bead of acoustical sealant is applied between layers and around anchor bolts.

PREP THE PLATE. To locate the bolts accurately, the mudsill is laid on edge across the top of the stemwall, and each bolt location is scribed onto the face of the 2×6.

NARROW WALLS REQUIRE OFFSET STRINGS. The tops of these stemwalls are only wide enough to carry the 2×6 walls, so the carpenters attach 2× spacer blocks to the stemwalls and then they fasten an offset stringline to the blocks to use as a reference for measuring.

DRILL THE LAYERS AS A SANDWICH. Although they started out marking and boring through each layer separately, the carpenters quickly learned that it's faster to stack up the poly, termite shield, and 2×6 mudsill, clamp them together, and drill through everything at once.

THE GASKET IS TREATED SEPARATELY. The soft and stretchy EPDM gasket (see the sidebar on p. 245) tends to get snagged and wrapped up by a spinning drill bit, so after the other layers are drilled the gasket is stapled to the underside of the mudsill and sliced with a utility knife at each bolt-hole location.

WHAT IT TAKES TO BE A PASSIVE HOUSE

THE PASSIVE HOUSE STANDARD AIMS TO maximize passive energy gains while minimizing energy losses. This is achieved with superinsulation, high-performance windows and doors, minimal thermal bridging, strict airtightness of the building envelope, mechanical ventilation, and optimal passive-solar gain. To attain Passive House certification, all of the building components are individually scrutinized in the Passive House Planning Package (PHPP), an elaborate spreadsheet program, before any construction begins. The PHPP predicts the performance of the house before it's built. Once built, the house is tested by a third party to ensure that it has achieved three performance requirements.

Air Infiltration

About 7 air changes per hour (ACH)

Code

≤0.6 ACH

Passive

Comparison: The IRC's current energy codes require houses to have no more than 7 air changes per hour (ACH) at 50 Pa.

BTU Consumption

About 47,550 BTU per sq. ft. annually

≤4,755 BTU per sq. ft. annually

Comparison: That's roughly 90% less heating and cooling than is required in a similarly sized code-built house.

Energy Usage

About 22 kwh per sq. ft.

≤11.1 kwh per sq. ft.

Comparison: This number, which includes heating, cooling, and electricity, is roughly half that of a typical house.

STARTS GOOEY AND STAYS THAT WAY. The primary air-seal in this assembly is Tremco® acoustical sealant. Highly elastic and sticky right out of the tube, this sealant won't harden over time like construction adhesives, so it creates a reliable air-seal at vulnerable joints.

PLASTIC COMES FIRST. After applying a thick, continuous bead of acoustical sealant to the top of the concrete, the carpenters lay the poly vapor retarder in place. They use hand pressure to push it firmly into the bead of sealant.

TERMITE SHIELD. Another bead of acoustical sealant is laid across the top of the poly before the termite shield, a copper-polymer composite membrane called YorkShield 106 TS (www.york-mfg.com), is placed over it.

SEAL THE JOINTS. Before placing the next 2×6 sill, a thick bead of sealant is applied to the edge of any adjoining sill. This is a commonly overlooked weak spot in an air-sealed assembly.

BELT AND SUSPENDERS. After all of the layers are in place and the foundation bolts have been fully tightened, another bead of sealant is applied to the exterior joint between mudsill and stemwall and at all butt joints.

GASKETS

Traditionally, the mudsill is laid atop a ¼-in.-thick polyethylene gasket. Although this sill sealer does help to reduce air leakage between the sill and the concrete, it's far from airtight. On this house, the builders installed a soft rubber EPDM gasket (BG63) made by Conservation Technology (www.conservationtechnology.com). Unlike with poly gaskets, the manufacturer claims that its EPDM gaskets will stay flexible at extremely low temperatures and will respond well to shrinkage and swelling even after decades of compression.

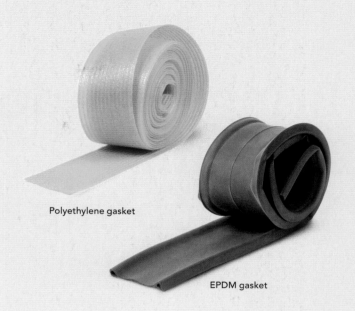

Polyethylene gasket

EPDM gasket

This part of the build typically is done on the carpenters' first day, so it's often their first hands-on involvement with the extreme airtightness requirement of this kind of house. In most cases, the carpenters never will have built even close to a Passive House level of airtightness, so establishing a good mental standard for the job starts here.

The mudsill is a one-shot deal. This project relies on several blower-door tests to evaluate air leakage, but the first test won't happen until the walls and roof are in place and sheathed. By then, any air leakage at the mudsill is far more difficult to address. It needs to be right the first time; there is no second chance here.

Learning to Love Acoustical Sealant

There are various sealants, gaskets, self-adhering membranes, and building tapes for air-sealing mudsills. Although we used a gasket in one layer of the mudsill assembly on this house—a belt-and-suspenders approach—most of the airtightness hinges on the use of Tremco acoustical sealant. Sold in tubes at specialty retailers and online, the black sealant installs easily with a caulk gun. It's exceed-

ingly sticky and highly elastic, and unlike construction adhesive, it never cures. Although the gooey, get-everywhere sealant makes for an interesting job site (you'll want to keep a large bottle of Goof Off® or Goo Gone® on hand), it is the most effective air-sealing solution I have found.

One of the issues I have in sealing mudsills with a rubber gasket alone is the treatment of butt joints and changes in direction. A healthy bead of sealant eliminates any concern about gaps in these areas.

Every Change Has Implications

In a Passive House, nailing down all of the building details before any principle construction work begins is exceedingly important. But even the best-laid plans are going to need last-minute tweaks. Here, the client feared that termites might move into the walls, where the double-stud framing would make it especially difficult to notice the infestation. In an effort to ease the client's mind while keeping with the builder's schedule, we decided to add a copper termite shield to the mudsill assembly. This termite shield later was trimmed back on the inside and covered with foam, eliminating the chance of a thermal bridge.

Bring Advanced Framing to Your Job Site

BY DANNY KELLY

Most houses have a lot more lumber in the walls than is really needed. All of that extra wood not only increases the costs, but also adds to thermal bridging and steals room from insulation.

Advanced framing aims to eliminate any lumber that isn't critical to the structure. Green-building programs award points for using advanced-framing techniques, which is great, but that's not why my firm does it. We do it because it allows us to use 2×6 studs and to install more insulation for about the same price as 2×4 walls. Also, the Department of Energy says a home with advanced framing will cost 5 percent less to heat and cool, which is a lot of money over the life of the structure.

Our path to advanced framing was incremental. We started about five years ago, when we began eliminating redundant jacks and cripples. Then we switched to 24-in. centers, two-stud corners, and ladder blocking at interior-wall intersections. These simple steps reduced the number of studs in the walls by 50 percent.

Once my crew felt comfortable with these changes, we made a switch to single top plates. This requires the framing members to be stacked within 1 in. of each other, creating an uninterrupted struc-tural load path from the roof to the foundation. We look for a corner where we can stack from the mudsill to the rafters, and we start our joist, stud, and rafter layouts there. One potential problem with using single top plates is that precut studs create a wall that's 1½ in. shorter than typical. This isn't a big deal with 8-ft. walls, because you can simply buy 8-ft. studs and cut them down, but with 9-ft. walls you have to cut down 10-ft. studs, which generates a lot of waste. Our solution is to cut the drywall a little shorter and use standard precut 9-ft. studs.

When the engineer allows, we also eliminate the conventional headers and use the band joist as a header. If the band joist alone is structurally insuffi-cient, we install a header between the band joist and the floor joists, which allows more insulation in the wall. If a header must go in the wall, we cover it with rigid foam on the exterior and use header hangers to eliminate jack studs.

We've found only one real drawback to advanced framing: With less lumber in the wall, there are fewer places to mount outlets, switches, and cabi-nets, so we often have to add blocking. It's impor-tant to communicate with your subs about this early so that these details can be worked out ahead of time.

INSTALL NAILERS FOR WINDOWS. Add 2×2s to the king studs to provide nailing for windows, siding, and exterior casing. Biscuits can reinforce corners on wide casing. Depending on the interior trim, you may have to do this for interior casing as well.

LONG WALLS ARE HARD TO LIFT. Advanced-framed walls are wobbly compared to conventionally framed walls. Even though they're lighter, you'll still need extra hands on deck. Small crews can frame walls in shorter segments to keep them under control as they're raised.

ELIMINATE EXTRA STUDS. When possible, eliminate extra jack and cripple studs by locating windows and doors so that at least one side of the opening falls within the normal stud layout. Be sure to leave room for casing, and consider the final elevation when moving doors and windows.

PICK A CORNER. Start the layout for the stacked framing at a corner where all the framing elements can be stacked from the mudsill to the rafters. An advantage of using a single top plate is that the stud layout on the top plate also can be used for joists or rafters.

BRACE WALLS WELL. As with conventional framing, begin plumbing and lining walls at corners, then at intersections. Unlike with conventional framing, you may need extra bracing to make up for more wobbly walls.

FASTEN DRYWALL AT INSIDE CORNERS WITH LADDER TEES AT INTERSECTING WALLS. Ladder tees reduce thermal bridging and use up small pieces of lumber that might otherwise go to waste. Locate the tees so that they line up with trim elements and drywall seams.

INVEST IN A CONNECTOR NAILER. In lieu of a second plate, intersecting wall sections are joined with metal tie plates. Each plate requires 32 nails. Assuming each fastener takes several seconds to hand-nail, this nailer paid for itself a long time ago.

STRING FOR STRAIGHT WALLS. Without a second top plate, advanced-framed walls built with warped lumber will look wavy, so use the straightest lumber possible. Stringlines and bracing can help to correct any problems.

USE HEADER HANGERS. In areas where you need conventional headers, use steel header hangers to eliminate jack studs.

STACKED FRAMING TRANSFERS THE LOAD. Aligned studs, joists, and rafters allow you to eliminate the second top plate in framed walls. Stacked framing also makes it easier to rough in pipes and ducts.

MOVE HEADERS INTO THE FLOOR FRAME. Window and door headers generally aren't needed in gable-end walls. Where headers are needed, the band joists (or engineered rim board) often can satisfy header requirements, as over these gable-wall windows.

SKIP HEADERS WHEN POSSIBLE. Openings in non-load-bearing walls don't need headers. In spots where you do need them, size them appropriately, and leave as much room as possible for insulation by using hangers or by moving the header into the floor above.

BE ON THE LOOKOUT FOR PROBLEMS. The author walks the site nearly every day. He marks unnecessary framing members for removal and indicates where additional blocking is needed.

Double-Stud Walls

BY RACHEL WAGNER

Energy-conscious builders pioneered double-stud walls after the oil-price shocks of the 1970s. This relatively low-tech method of building energy-efficient walls uses common materials and familiar assemblies. These walls have several benefits in addition to their high R-value: Thick cellulose-insulated walls are quiet, and many homeowners appreciate the deep window stools. In addition, the framing method virtually eliminates thermal bridging within the wall assembly, although there still can be thermal bridges at sills, top plates, and window and door openings.

The basic strategy is simple: An exterior wall is built from two parallel stud walls. Both stud walls and the space between them are filled with continuous insulation. The exterior is sheathed and finished conventionally, although a rain-screen siding detail is recommended. Of course, there are some important design and construction considerations, starting with the appropriate thickness.

One Size Does Not Fit All

You can vary the thickness of double-stud walls for each project or climate to achieve an overall R-value that fits. The R-value of dense-packed cellulose (the most common insulation used in double-stud walls) is about 3.7 or 3.8 per inch, so a 12-in. double-stud wall has an R-value of about 45.

Both the design of the building and your performance goals affect the optimum thickness of your walls. The walls might be 16 in. thick for a Passive House in Vermont, but only 10 in. thick for a low-energy house in Iowa. A bigger or more complicated building usually warrants thicker walls (with a higher R-value) than a small, simple building.

Wall construction can vary depending on the type of foundation, the type of floor system, and the preference of the builder or designer. For a slab on grade, a 2×6 outer wall allows the outer stud wall to be situated so that the framing bears on the slab and also extends past the slab edge to cover the vertical rigid insulation at the slab. With a basement, the double-stud wall sits on the floor framing, so both walls usually can be framed with 2×4s. Studs can be either 16 in. or 24 in. on center; be sure that the interior finish materials and the siding are compatible with the stud spacing.

One Sill Plate or Two?

Separate stud walls with individual sill plates will be easier to construct and more energy efficient, but a shared bottom plate can be useful when framing on

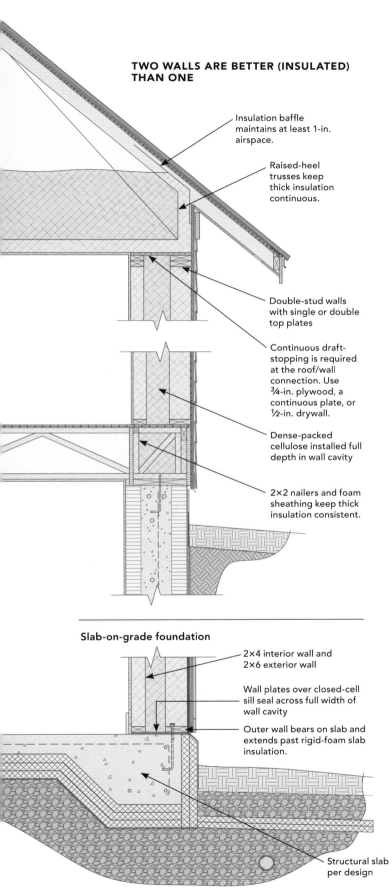

TWO WALLS ARE BETTER (INSULATED) THAN ONE

Insulation baffle maintains at least 1-in. airspace.

Raised-heel trusses keep thick insulation continuous.

Double-stud walls with single or double top plates

Continuous draft-stopping is required at the roof/wall connection. Use ¾-in. plywood, a continuous plate, or ½-in. drywall.

Dense-packed cellulose installed full depth in wall cavity

2×2 nailers and foam sheathing keep thick insulation consistent.

Slab-on-grade foundation

2×4 interior wall and 2×6 exterior wall

Wall plates over closed-cell sill seal across full width of wall cavity

Outer wall bears on slab and extends past rigid-foam slab insulation.

Structural slab per design

an insulated slab foundation. A shared top plate isn't required, but installing a continuous ¾-in. plywood top plate can meet fire-blocking requirements and make it easier to install floor or roof trusses.

Most framers erect the outer wall separately and first; this approach most resembles conventional stud-wall construction. When framing the inner wall, it is important to align window and door openings perfectly; the openings will be joined later using plywood box frames, like big gussets. The studs in the two walls can be either parallel or staggered, but they must line up at the openings. Parallel studs are nice for insulation netting and draft-stopping; of course, the studs won't be parallel at the corners. Some builders like to construct double-stud wall trusses using plywood gussets; in this case, the walls go up together.

Air-Sealing Still Matters

Lots of insulation does not guarantee better energy performance. It's also essential to minimize air leaks within wall construction. This enhances energy performance and reduces the risk of moisture intrusion into the wall.

Be sure to define the interior air barrier, which can be drywall if you're using the airtight-drywall approach, a product such as MemBrain™ (www.certainteed.com), a layer of OSB under the drywall, or even polyethylene if you're building in a very cold climate. Continuously seal all joints and connections at seams, corners, floors, ceilings, and window and door openings. Seal around all electrical boxes, wire and conduit, and duct penetrations.

PERFORMANCE REQUIREMENTS FOR DOUBLE-STUD WALLS

FIRE BLOCKING

The International Residential Code (IRC section R302.11) requires draft-stopping in double-stud assemblies every 10 ft. (minimum) along the length of the wall, from bottom plate to top plate and covering the full depth of the double cavity, using 1/2-in. gypsum drywall or 3/4-in. plywood. (Drywall is easier to cut and fit into place.) The code also requires fire blocking to keep the top of the wall assembly separate from the floor framing or attic spaces above. If you're not using a full-depth top plate that spans across both stud walls, install 1/2-in. drywall or 3/4-in. plywood between the top plates, and fire-caulk the joints.

INSULATION

Current practices favor dense-packed cellulose or fiberglass blown through fabric into the wall. With thick cavities not fully divided into neat individual bays, it's important to maintain the required density of the insulation to prevent it from settling, which would leave an uninsulated gap at the top of the assembly. Some builders "net" each bay, fastening filter fabric across the depth of the two studs essentially to create individual full bays. Then they fasten the fabric across the front of the wall assembly, as is typical for dense-packing, and blow the insulation into each bay as they would for a single-stud wall. (If you want to use this technique, you must align the studs in both walls.) Another technique uses the horizontal draft-stopping as the containment for the insulation, although it is placed at intervals of 8 ft. on center instead of 10 ft. Filter fabric is then used only at the face of the framing.

MOISTURE

A double-stud wall slows heat loss from the building better than a single-stud wall, so the exterior sheathing will be colder and potentially wetter in winter than it would be in a typical single-stud wall. In most climates where a double-stud wall will be used, the code requires a vapor retarder on the warm side of the wall; vapor-retarder paint can satisfy this requirement. Plywood or structural fiberboard sheathing will give the wall a better chance to dry outward than OSB, and installing the siding over furring strips also helps the sheathing to stay dry.

Framing for Efficiency

BY STEVE BACZEK

The frame of a Passive House may not be as exciting as the thick layers of insulation, the high-tech mechanical systems, or the triple-glazed windows, but it plays a very important supporting role—pun intended—in achieving success.

I chose every component of the framing package in this house with care, and for a specific reason. The exterior sheathing provides airtightness, the double-stud walls and raised-heel roof trusses are a cost-effective means of supporting or containing above-average levels of insulation, and the floor trusses easily span the open floor plan and provide plenty of room for the many ducts necessary for the ventilation system and supporting mechanicals.

Two Walls, Two Air Barriers

The chief function of the double-stud walls is to hold insulation. Measuring 14 in. from the interior face of the 2×4 inner wall to the exterior face of the 2×6 outer wall, the wall assembly provides a 5-in. thermal break between halves.

In addition to the taped sheathing seams and the picture-frame-style application of Tremco acoustical sealant, the wall is redundantly air-sealed with a 4-in.-thick coat of closed-cell foam sprayed against the inside face of the sheathing. The remainder of the cavity is filled with 10 in. of dry, dense-packed cellulose, bringing the wall assembly to an overall thermal resistance of about R-52.

Some energy-conscious builders and architects might wonder why anybody would design a building that represents the height of energy efficiency, and specify walls with 2×6 studs spaced 16 in. on center

THE INTERIOR FRAME provided the typical mounting surface for fixtures and finishes, but it also served as a barrier to contain the cavity insulation before the final wallboard was installed.

MULTIPURPOSE COMPONENTS

BECAUSE THE ATTIC WON'T BE USED FOR ANYTHING BUT insulation, raised-heel roof trusses were a quick and cost-efficient means of putting a lid on this house and of providing a way to hang and finish the ceiling air barrier before partition walls went in.

OPEN-WEB FLOOR TRUSSES offer long spans and the ability to run mechanicals between the first and second floors—a saving grace in a house where the attic is off-limits and there is no basement.

THE OUTSIDE FRAME supports most of the floor load and all of the roof load. It also holds part of the primary air barrier, and with the help of diagonal bracing, provides the necessary wind-shear resistance for this coastal site.

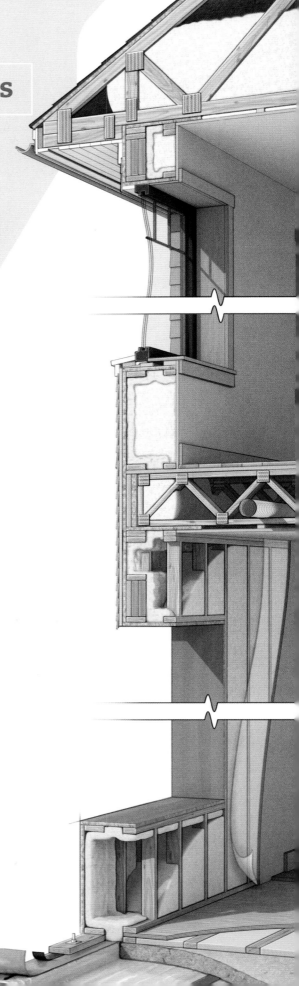

and structural headers over every window—details that fly in the face of advanced-framing techniques.

Given that these walls are thermally broken by the cavity between the inner and outer stud walls, the only advantage to framing with studs set on 24-in. centers would be a small cost savings in lumber. But on this house, which is located in a coastal high-wind zone, that small savings would have been offset by the additional structural measures required of a wall 24 in. on center.

For the headers, all of which are thermally broken with a piece of 2-in.-thick rigid foam, I have found that consistency pays even if it results in a minor energy penalty. I try to minimize decisions for the builder to increase the likelihood that the things I need to be done right will be done right. Also, for what it's worth, the high-performance triple-glazed windows necessary in a Passive House are two to three times the weight of a typical double-glazed window, so a robust window frame is a good thing.

At the top of each wall, bridging the gap between the interior and exterior stud walls, is a rip of ¾-in. plywood, which has a couple of duties. First, it caps and isolates the cavity space of the double-stud wall. Second, it overhangs the interior wall plate, providing a means to connect the interior ceiling to the wall assembly, maintaining the continuity of the air barrier. A third function is one I hadn't planned for but that the builders found very useful: a walking surface. By attaching bracing below the plywood flange, the builders were able to walk the walls easily while installing the floor and roof trusses.

Strong, Wide-Open Floor Framing

One of my goals in designing a successful Passive House is to get as much of the structural load from the floors and roof as possible to the outside of the house. This allows me to keep an open floor plan, which is helpful in moving conditioned air around the house. To achieve this open floor plan, I needed engineered floor joists, which can span longer distances than dimensional lumber.

Although it's largely heated by the sun in the winter, a Passive House still relies on mechanical systems. In addition to the standard plumbing and electrical, this house has lots of ductwork for ventilation. Without a basement or conditioned attic, just about everything has to run through the floor joists that support the second level.

This combined need for a long span and room for lots of mechanicals made open-web floor trusses an easy choice. They are cost-effective and sturdy, and they eliminate concerns about the placement of penetrations or the need for mechanical chases.

This Attic Is for Insulation

Because the attic will hold 24 in. of loose-fill cellulose, I didn't even attempt to provide storage or living space up there. Forfeiting any claim to the attic made roof trusses an easy choice compared to a traditional stick-built roof, and allowed me to get the house dried in and prepped for the uppermost portion of the primary air barrier: the ceilings.

To ensure that this ceiling air barrier—a layer of veneer-plastered blueboard—would be continuous, the plaster installers hung and finished the ceiling before any interior partition walls were framed. That approach not only eliminated the hundreds of linear feet of joints between top plates and ceiling joists—all of which are weak points in an air barrier—but it made the hanger's job easier because full sheets of blueboard could be used.

THE CONSTRUCTION SEQUENCE IS GUIDED BY BLOWER-DOOR TESTS

THE AIRTIGHTNESS REQUIREMENT FOR PASSIVE HOUSE CERTIFICATION IS LESS than or equal to 0.6 air changes per hour at –50 Pa (ACH50). This number can also be expressed as 177 cu. ft. per minute at –50 Pa (cfm50). I prefer to use the cfm figures because the larger number makes any changes in performance easier to track.

TEST RESULT: 177 CFM50

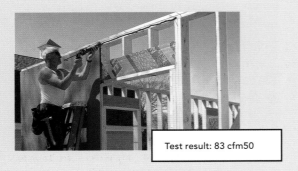

Test result: 83 cfm50

Test result: 46 cfm50

PHASE 1: PRIMARY AIR BARRIER. The primary air barrier in this house is formed by the slab, the exterior sheathing, and the plastered ceilings below each roofline. Backed up with a thick, continuous bead of Tremco acoustical sealant at all seam edges, the 7/16-in.-thick ZIP System® OSB was chosen because of its butyl-based seam-sealing tape, which partners with the water- and air-resistive barrier that's bonded to the exterior side of each sheet. The rough openings at windows and doors are left uncut so that the builders can test the airtightness of the shell before moving to the next step.

PHASE 2: SECONDARY AIR BARRIER. A 4-in.-thick coat of closed-cell polyurethane-foam insulation is sprayed on the inner face of the wall sheathing. Running continuously from the subslab insulation to the top plates, the foam performs a couple of functions. First, it acts as a secondary air barrier should there be any air leakage through the sheathing seams or the acoustical sealant. Second, with an insulation value of roughly R-27, it ensures that the inner surface of the sheathing remains above the dewpoint.

Test result: 106 cfm50

Test result: 110 cfm50

PHASE 3: WINDOWS AND DOORS. Even a house with the best windows and doors in the world, installed perfectly, is leakier than a house without any openings. For that reason, once the sheathing has been cut away at rough openings and the windows and doors are in place, the air-leakage numbers will creep up slightly. Also, because most penetrations should have been made at this point, this test result should be a fairly accurate prediction of the final result once the house is complete.

PHASE 4: INSULATION AND MECHANICALS. With the more-delicate control functions handled as close to the outside face of the building as possible, the space between the outer and inner stud walls and the space in the attic offer an opportunity for more-cost-effective insulation to provide the bulk of the thermal resistance. Cellulose insulation is packed into the walls to a density of 3.6 lb. per sq. ft., adding R-37 to the overall thermal resistance of the wall assembly. In the attic, loose-fill cellulose is piled to a depth of 24 in., providing an R-92 insulated lid above the ceiling air barrier of the house.

Jim Anderson is a framing contractor in Littleton, Colorado.

Rick Arnold is a veteran builder and contributing editor to *Fine Homebuilding*. He lives and works in North Kingstown, Rhode Island.

Steve Baczek is an architect in Reading, Massachusetts. Construction for both "Air-Sealed Mudsill Assembly" and "Framing for Efficiency" by Dunhill Builders, Osterville, Massachusetts (www.dunhillbuilders.com).

Don Burgard is senior copy/production editor at *Fine Homebuilding*.

Michael Chandler owns Chandler Design-Build (www.chandlerdesignbuild.com) near Chapel Hill, North Carolina.

Bruce Coldham, FAIA, is a principal of Coldham & Hartman Architects (www.coldhamandhartman.com) in Amherst, Massachusetts.

Clayton DeKorne is the author of *Trim Carpentry and Built-Ins* (The Taunton Press, 2002). A third-generation carpenter, he is currently executive editor of the *Journal of Light Construction*. He lives and works in Brooklyn, New York.

Chris Ermides is a former *Fine Homebuilding* associate editor. In addition to writing and carpentry, he fills his time as a freelance web producer for Taunton's Workshop e-learning series. He lives in Saratoga Springs, New York, with his wife and two sons.

Justin Fink is the Project House manager at *Fine Homebuilding*.

Joseph Fratello is an electrical contractor based in Southampton, New York.

David Grandpré is a registered professional engineer with C.A. Pretzer Associates, Inc., in Cranston, Rhode Island.

Mike Guertin, author of *Precision Framing* (The Taunton Press, 2001), is a builder, remodeler, and editorial adviser to *Fine Homebuilding*.

Larry Haun, author of *The Very Efficient Carpenter* (The Taunton Press, 1992) and *Habitat for Humanity How to Build a House* (The Taunton Press, 2008), framed houses for more than 50 years. Donations in Larry's memory can be made to Habitat for Humanity.

Ryan Hawks is a carpenter who does side jobs in San Diego, California.

Lynn Hayward owns and operates Hayward-Robertson Builders in Northport, Maine.

Danny Kelly began framing houses in 1989. He and business partner Dan McArdle own Kelly McArdle Construction in Charlotte, North Carolina. When Danny isn't building houses, he's building couch forts with Danny III, constructing train villages with twins Tommy and Jimmy, or starting but never finishing multiple projects for his wife, Molly.

Patrick McCombe is an associate editor at *Fine Homebuilding*.

Rob Munach, P.E., operates Excel Engineering in Carrboro, North Carolina.

Mike Norton lives in Middleboro, Massachusetts. His website is www.frametechs.com.

Roe A. Osborn, former *Fine Homebuilding* editor, is an artist, writer, and photographer on Cape Cod, Massachusetts. His chapter was adapted from his book *Framing a House* (The Taunton Press, 2010).

Bryan Readling, P.E., is an engineer with APA-The Engineered Wood Association's Field Services Division in Davidson, North Carolina.

Fernando Pagés Ruis, builder with 30 years of experience, is the author of *Building an Affordable House* (The Taunton Press, 2005) and *Affordable Remodel* (The Taunton Press, 2007).

Brian Saluk is a custom home-builder from Berlin, Connecticut.

Debra Judge Silber is the managing editor of *Fine Homebuilding*.

John Spier is a builder on Block Island, Rhode Island, and the author of *Building with Engineered Lumber* (The Taunton Press, 2006).

Michael Springer is a tool tester and industry journalist and consultant in Boulder County, Colorado.

David Utterback is a former builder and a certified building inspector who conducts seminars on building codes and wood-frame construction.

Rachel Wagner is a principal at Wagner Zaun Architecture (www.wagnerzaun.com) in Duluth, Minnesota.

Rob Yagid is the executive editor of *Fine Homebuilding*.

All photos are courtesy of *Fine Homebuilding* magazine © The Taunton Press, Inc., except as noted below:

The articles in this book appeared in the following issues of *Fine Homebuilding*:

pp. 5–10: Choosing the Right Framing Nailer by Michael Springer, issue 230. Photos by Rodney Diaz except for photo p. 5 by Justin Fink and left photo p. 8 and bottom photos p. 9 courtesy of Bostitch.

pp. 11–12: What's the Difference: Hammers by Rob Yagid, issue 205. Photos by Rodney Diaz.

pp. 13–14: What's the Difference: Framing lumber by Don Burgard, issue 222. Photo by Rodney Diaz.

pp. 15–17: How it Works: Nails by Debra Judge Silber, issue 245. Photo by Justin Fink. Drawings by Christopher Mills.

pp. 18–21: Building Skills: Building sturdy sawhorses by Patrick McCombe, issue 218. Photos by Rob Yagid, except for bottom left photo p. 19.

pp. 22–29: Exploring the Benefits of Engineered Floor Joists by Chris Ermides, issue 209. Photos by Dan Thornton except for photos p. 22 by Justin Fink and bottom photo p. 24 by Steve

Culpepper. Truss tags pp. 26–27 courtesy of Structural Building Components Association and the Truss Plate Institute. Drawings by Dan Thorton except for drawings pp. 26–27 by Bill Godfrey.

pp. 30–34: Managing Job-Site Mud by Fernando Pages Ruiz, issue 217. Photos by Dan Thornton except for photo p. 30 by Roe A. Osborn. Drawing by Christopher Mills.

pp. 35–40: Temporary Power on the Jobsite by Joseph Fratello, issue 229. Photos by Charles Bickford except for top photo p. 36 courtesy of American Honda Motor Co. and bottom photo p. 36 courtesy of APC/Schneider Electric. Drawings by Dan Thornton.

pp. 42–49: 10 Rules for Framing by Larry Haun, issue 158. Drawings by Christopher Clapp.

pp. 50–52: Anchoring Wood to a Steel I-Beam by John Spier, issue 180. Photos by Krysta S. Doerfler. Drawings by Bob La Pointe.

pp. 53–55: How it Works: Wall Framing by Rob Munach, issue 214. Photos by Daniel S. Morrison. Drawing by Don Mannes.

pp. 56–63: Framing with a Crane by Jim Anderson, issue 140. Photos by Mike Rogers except for photo p. 57 by Andy Engel and bottom right photo p. 58, top photo p. 59, and top photos p. 61 by Jim Anderson.

pp. 64–71: All about Headers by Clayton DeKorne, issue 162. Photos by Scott Phillips except for photo p. 64 by Roe A. Osborn. Drawings by Dan Thornton.

pp. 72–79: Common Engineering Problems in Frame Construction by David Utterback, issue 128. Photos by Western Wood Products Association. Drawings by Christopher Clapp.

pp. 80–-86: Wind-Resistant Framing Techniques by Bryan Readling, issue 237. Photos by Brian Readling except for photos p. 84 by Patrick McCombe. Drawings by Vince Babak.

pp. 87–93: Where Do You Want the Blocking by Justin Fink, issue 190. Photo p. 87 by Rob Yagid, top left and bottom photos p. 88 by Roe A. Osborn, top right photo p. 88 by Justin Fink, center left photo p. 88 by Tom Meehan, and center right photo p. 88 by Brian Pontolilo. Drawings by Bruce Morser.

pp. 95–102: The Well-Framed Floor by Jim Anderson, issue 160. Photos by Chris Green. Drawing by Chuck Lockhart.

pp. 103–108: Fast, Accurate Floor Sheathing by Danny Kelly, issue 229. Photos by Patrick McCombe. Drawing by Don Mannes.

pp. 109–116: Built-Up Center Beams by Rick Arnold and Mike Guertin, issue 144. Photos by Roe A. Osborn except for bottom photo p. 113 by Tom O'Brien. Drawing by Dan Thornton.

pp. 117–124: Frame a Strong Stable Floor with I-Joists by John Spier, issue 197. Photos by Justin Fink except for bottom photo p. 124 by Krysta S. Doerfler. Drawings by Bob La Pointe.

pp. 125–130: Supporting a Cantilevered Bay by Mike Guertin, issue 136. Photos by Roe A. Osborn.

pp. 131–140: 6 Ways to Stiffen a Bouncy Floor by Mike Guertin and David Grandpré, issue 184. Photos by Mike Guertin except for photos p. 131 by Daniel S. Morrison, left photo p. 135 by Justin Fink, and photos p. 138 and 140 by Krysta S. Doerfler. Drawings by Don Mannes.

pp. 142–149: Careful Layout for Perfect Walls by John Spier, issue 156. Photos by Roe A. Osborn. Drawings by Vince Babak.

pp. 150–157: Fast and Accurate Wall Framing by Mike Norton, issue 242. Photos by Charles Bickford.

pp. 158–166: Laying Out and Detailing Wall Plates by Larry Haun, issue 126. Photos by Andy Engel

except for the top photo p. 164 by Larry Haun. Drawings by Rick Daskam.

pp. 167–172: Not-So-Rough Openings by John Spier, issue 176. Photos by Roe A. Osborn. Drawings by Dan Thornton.

pp. 173–180: Framing Curved Walls by Ryan Hawks, issue 148. Photos by Tom O'Brien. Drawings by Dan Thornton.

pp. 181–189: Framing Big Gable Walls Safely and Efficiently by Lynn Hayward, issue 181. Photos by Daniel S. Morrison. Drawings by Heather Lambert.

pp. 190–198: Raising a Gable Wall by John Spier, issue 122. Photos by Roe A. Osborn.

pp. 199–203: Better Framing with Factory-Built Walls by Fernando Pagés Ruiz, issue 169. Photos by Fernando Pagés Ruiz.

pp. 204–205: How it Works: Shear walls by Rob Yagid, issue 222. Photo courtesy of FEMA/Andrea Booher. Drawings by Don Mannes.

pp. 206–209: A Slick Approach to Straightening Walls by Roe A. Osborn, issue 214. Photos by Roe A. Osborn. Photo illustration by Bill Godfrey.

pp. 211–213: Curved Ceiling? No Problem by Michael Chandler, issue 185. Photo by Seth Tice-Lewis. Drawings by Toby Welles.

pp. 214–222: Framing Cathedral Ceilings by Brian Saluk, issue 118. Photos by Greg Morley except for photo p. 214 by Scott Gibson. Models by Linden Frederick.

pp. 223–228: Open Up the Ceiling with a Steel Sandwich by Michael Chandler, issue 178. Photos by Seth Tice-Lewis. Drawings by Heather Lambert.

pp. 230–240: Six Proven Ways to Build Energy-Smart Walls by Bruce Coldham, issue 208. Drawings by John Hartman.

pp. 241–245: Air-sealed Mudsill Assembly by Steve Baczek, issue 241. Photos by Justin Fink except for product photos by Dan Thornton. Drawings by John Hartman.

pp. 246–251: Bring Advanced Framing to Your Job Site by Danny Kelly, issue 226. Photos by Patrick McCombe.

pp. 252–254: Energy Smart Details: Double-stud walls by Rachel Wagner, issue 228. Drawings by Elden Lindamood, courtesy of Wagner Zaun Architecture.

pp. 255–258: Framing for Efficiency by Steve Baczek, issue 244. Photos by Justin Fink. Drawing by John Hartman.